T0328640

Industrial Process Plant Construction Estimating and Man-Hour Analysis

Industrial Process Plant Construction Estimating and Man-Hour Analysis

Kenneth Storm

Gulf Professional Publishing
An imprint of Elsevier

Gulf Professional Publishing is an imprint of Elsevier
50 Hampshire Street, 5th Floor, Cambridge, MA 02139, United States
The Boulevard, Langford Lane, Kidlington, Oxford, OX5 1GB, United Kingdom

Notices
Knowledge and best practice in this field are constantly changing. As new research and experience
broaden our understanding, changes in research methods, professional practices, or medical
treatment may become necessary.

Practitioners and researchers must always rely on their own experience and knowledge in evaluating
and using any information, methods, compounds, or experiments described herein. In using such
information or methods they should be mindful of their own safety and the safety of others, including
parties for whom they have a professional responsibility.

To the fullest extent of the law, neither the Publisher nor the authors, contributors, or editors, assume
any liability for any injury and/or damage to persons or property as a matter of products liability,
negligence or otherwise, or from any use or operation of any methods, products, instructions, or
ideas contained in the material herein.

Library of Congress Cataloging-in-Publication Data
A catalog record for this book is available from the Library of Congress

British Library Cataloguing-in-Publication Data
A catalogue record for this book is available from the British Library

ISBN: 978-0-12-818648-0

For information on all Gulf Professional publications visit
our website at https://www.elsevier.com/books-and-journals

Publisher: Brian Romer
Senior Acquisition Editor: Katie Hammon
Editorial Project Manager: Michelle W. Fisher
Production Project Manager: Anitha Sivaraj
Cover Designer: Matthew Limbert

Typeset by SPi Global, India

Working together
to grow libraries in
developing countries

www.elsevier.com • www.bookaid.org

Contents

12. Statistical applications to construction

Preface

Construction projects are divided into three sectors: building, infrastructure, and industrial construction.

Building construction: This sector is divided into residential and commercial building.

Infrastructure: Infrastructure is divided into heavy civil engineering that includes airports, bridges, dams, tunnels, highways, hydropower, water treatment, distribution, and rapid-transit systems.

Industrial construction: Industrial construction includes refineries, power generation, mills, and manufacturing plants. These projects are identified by the plant process and vary in size and complexity.

This edition **of Industrial Process Plant Construction Estimating and Man-Hour Analysis** focuses on industrial process plants and enables the estimator to apply statistical applications, estimate data tables, and estimate sheets to use methods for collecting, organizing, summarizing, presenting, and analyzing historical man-hour data. As construction processes become more complex, contractors are placing a greater importance on specialized education. It is increasingly important for estimators, engineers, and construction managers to have a bachelor's degree in a construction-related field. A strong background in mathematics and statistics is essential. Estimators are required to set up cost codes, summarize and analyze historical data and trends, and maintain labor data bases by entering and backing up data. Construction-work estimating data sets or books are used for construction analysis and cost estimating. These data sets and books need to be evaluated for accuracy, reliability, verifiability, and convenience. The construction business requires statistical methods, indexing, cost analysis, and estimating to provide detailed accurate bid proposal to handle increasing competition. Engineers and contractors use statistical methods, historical data, and man-hour tables, which are entered into the computer, that model the work of construction. The purpose of this book is to provide the reader the latest estimating and statistical methods to evaluate the accuracy and verify historical data collected from field installation of process piping and equipment in industrial process plants. The book begins with the introduction devoted to labor, productivity measurement, collection of historical data, estimating methods, and factors affecting construction labor productivity and impacts of overtime. Then, estimating is developed, and project sections for equipment are given. Engineering, contractors, and owners are concerned with statistical applications to industrial process plants and this section is provided. The book develops the principals in a systematic way, and the book is divided into 12 chapters. The first chapter is an introduction to construction statistics using Excel templates and mathematical spreadsheets. The purpose of this chapter is to provide a basic understanding of how to use Excel templates and spreadsheets specific to construction. These Excel templates and spreadsheets are used to provide a step-by-step guide for how to set up and perform a wide variety of statistical applications to industrial process plants. Chapter 2 provides the reader the man-hour schedules and tables for the installation of process piping in industrial process plants. Chapters 3–10 provide detailed scopes of work for each section and man-hour tables based on historical data collected from field erection of equipment installed in industrial process plants. Estimate sheets that itemize the equipment scope of work are set up on a computer, and historical man-hours, using the unit quantity model, calculate the direct craft man-hours. Installation man-hours are summarized, and historical man-hours are compared with estimated man-hours. Chapters 11 and 12 provide sample estimates and illustrations of statistical applications to industrial process plants. The direct craft man-hours have been verified by statistical analysis and were determined from time and measurement of craft labor for field erection of process piping, equipment, and boilers installed in process plants throughout the country. The sections' work scope, man-hour tables, and estimate sheets are applicable for all contractors, engineering firms, and process plants that install process piping, equipment, and boilers in industrial process plants. The statistical and estimating methods in this book will enable estimators and engineers to prepare comprehensive and detailed direct craft man-hour estimates, RFPs, and field change orders for the following industrial process plants: diesel, solar, and coal-fired power plants; industrial package boilers; pulp and paper plants; oil refinery; and boiler tube replacement.

Construction experience and education may vary, and we believe that the book is suitable for those working in engineering and construction. The book has been written to appeal to engineering/technology/construction estimating and management settings. The book will decrease the chance of error and allow estimators to accurately determine the actual direct craft man-hours for the complete installation of process piping, equipment, and boilers. The estimating methods in this book will enable the estimator to use the comparison method to estimate the differences between proposed and previously installed equipment and boilers, and the unit quantity method will be a final check on the estimated man-hours compared with the historical man-hours. The book does not include man-hours and cost for material, equipment usage, indirect craft, supervision, project staff, warehousing, and storage. The direct craft man-hour estimate is the basis for the estimator to obtain the project schedule and the cost for mobilization and demobilization, indirect craft, supervision, project staff, construction equipment, subcontractors, material, site general conditions, and overhead and fee. In addition, the estimator must determine all factors that will affect productivity and overtime impacts.

Review of the preface and introduction will enable the reader to understand craft labor productivity, productivity measurement, collection of historical data, estimating methods and labor factors, and loss due to labor productivity and overtime impacts

To apply the scopes of work, direct craft man-hour tables, and estimating sheets, the reader must be familiar with the following chapters entitled "Introduction and Chapters 1, 11, and 12." These chapters illustrate, with practical examples, the steps required to estimate and analyze man-hours to uncover the relationships that exist for process plant cost and labor:

(1) Introduction will enable the reader to set up cost codes and methods for collecting, organizing, summarizing, presenting, and analyzing historical man-hour data using graphic and regression analysis to verify historical data.

(2) Chapter 1 enables the reader to set up and use Excel templates and spreadsheets to automate statistical functions to perform mathematical and statistical applications to make the connection to process plant construction.

(3) Chapter 11 provides the reader the "industrial process plant construction estimating process" to enable the reader to use statistical and estimating methods, scopes of work, man-hour tables, and estimate sheets to the following:

   (a) Evaluate the accuracy and verify historical data collected for process piping and equipment installed in industrial process plants

   (b) Provide a comprehensive and accurate method using construction statistics and estimating methods to compile detailed craft man-hour estimates for bid proposals, RFPs, and field change orders

(4) Chapter 12 provides the reader the knowledge to use construction statistics to forecast, use learning curves and time series to validate data, and prepare detailed estimates. This chapter includes practical examples of statistical applications and methods to help the reader understand the importance of man-hour analysis and estimating with the intention to point out the connection to construction. Construction statistics depends on statistical and mathematical methods and is an important part of field cost and construction man-hour analysis. The book will be a source for those engaged in estimating, forecasting, managing, and bidding projects in the industrial construction industry.

Introduction

Introduction

This section provides methods for collecting, organizing, summarizing, presenting, and analyzing historical man-hour data. Labor productivity is one of the most important items in a construction estimate. To estimate the productivity for work, there is a dependence on the value of historical man-hours collected in field construction. This section, for the most part, provides methods to determine productivity measurement of construction labor. Detailed methods of estimating are based on historical data that have been collected, structured, and verified by mathematical and statistical analysis. Craft man-hour estimates are developed by using detailed construction-work estimates that apply to any complexity of design and use historical and quantitative data that lead to a cost driver easily understood. The purpose of this section is to provide the reader the basic understanding to set up cost codes, methods for collecting, organizing, summarizing, presenting, and analyzing historical man-hour data using graphic and regression analysis to verify historical data. Labor, productivity measurement, tracking systems, estimating methods, and factors for labor productivity and impacts of overtime are given to enable the estimator to prepare detailed accurate estimates. The historical man-hour data in this book have been verified by measurement, project cost reports from field erection, foremen's report, and one-cycle time studies, and the data are revised continuously due to construction design, engineering, labor skill, material, equipment, and procedures. Information for man-hour analysis is obtained from the foreman's report and is used to find the number of man-hours for a task and time control. The cost engineer and welding quality control in the field monitor and verify the work. These reports are collected for field construction work. From the reports and review of the specifications, codes and drawings the cost engineer and the estimator will examine the data for consistency, completeness, and accuracy. Reports are collected for similar work, and the data are entered into a spreadsheet. The spreadsheet prepares the data for mathematical analysis. The engineer and estimator determine the productivity rates. The rate is used for future cost analysis and estimating similar scopes of work. The estimate data are based on "standard," which is defined as "forming a basis for comparison." The standard unit man-hour involves these considerations: The work is being performed by a contractor who is familiar with all conditions at the job site; the project has the proper supervision; the workers are familiar with and skilled in the performance of the work task; and there is an adequate supply of labor. There are clarifications and exceptions stated for the application of the data.

Labor productivity and analysis

Labor—Man-hour unit rates

Labor productivity is concerned with direct craft labor. Direct craft labor means the craft is working in the field erection of process piping and equipment. The man-hour is defined as the amount of work performed by the average worker in one hour.

Formula: Man-Hour = Time × Quantity (Refer to the Unit Quantity Method)

Once time values are known for a construction task, they are multiplied by the quantity. Time may be for individuals or for crew work, and it is based on the construction task. Time is expressed relative to a unit of measure, such as LF, EA, SF, and Ton. The unit of time may be a minute, hour, day, month, or year.

Examples of man-hour units are as follows:

(1) HRSG—seal weld side, roof, and floor casing field seams; number of welder man-hours per lineal foot of field welding, 0.35 MH/LF
(2) Welding butt weld, carbon steel, arc-uphill, WT <= 0.375″; number of welder man-hours per diameter inch of welding, 0.50 MH/DI

The union/nonunion craft, experience factor, and PF&D allowances have been included in the craft unit man-hours. Standard unit rates can be used to estimate work anywhere in the United States. The estimator must determine all factors that will affect labor productivity and overtime impacts. The man-hour units and quantities are based on historical data that have been verified by statistical analysis, and the man-hour rates are competitive.

Productivity measurement

Historical records provided the direct craft man-hour data for field installation of piping and equipment. Two methods for the measurement of construction time were used to collect, analyze, and compile the actual man-hour data in this book.

(1) Foreman report—job cost by cost code and type
(2) Nonrepetitive one-cycle time study
(3) There are several ways in which actual time data are compiled and analyzed from the foreman's report. One method is to compile data for man-hour analysis obtained from the foreman's report for mathematical analysis. Then, the engineer will determine the productivity rates. The productivity rates are then entered into the estimating system to be used to estimate future work that is similar.

The following example will enable the reader to set up cost codes and analyze historical data, from field erection cost reports, using graphic analysis to verify direct craft man-hours.

Analysis of Foreman's Report
Coal-Fired Power Plant
Project: Boiler Erection-Field Welding Rates
JOB COSTS BY COST CODE AND COST TYPE

		HOURS PER DIEM
PHASE	DESCRIPTION	Tube welding—MH per weld
LABOR	BURDEN	Design pressure (PSI)
	Tube size (OD)	1501–2000
000000	Over 1″ and including 1-1/2″ heli arc	3.00
000000	Over 1-1/2″ and including 2″ heli arc	3.70
000000	Over 2″ and including 2-1/2″ heli arc	4.20
000000	Over 2-1/2″ and including 3″ heli arc	4.80
000000	Over 3″ and including 3-1/2″ heli arc	5.30
000000	Over 3-1/2″ and including 4″ heli arc	5.90
000000	Over 4″ and including 4-1/2″ heli arc	6.50
000000	Over 4-1/2″ and including 5-1/2″ heli arc	7.70
000000	Over 5-1/2″ and including 6-1/2″ heli arc	9.00

Foreman's report

Information for man-hour analysis is obtained from the foreman's report. The foreman's report is used to find the number of man-hours for a task. The report is used for cost and time control. The cost engineers and welding quality control in the field monitor and verify the work. These reports are collected for the field installed piping and equipment. From the reports and review of the specifications, codes, and drawings, the cost engineer and the estimator will examine the data for consistency, completeness, and accuracy. Reports are collected for similar work, and the data are entered into a spreadsheet. The spreadsheet prepares the data for mathematical analysis. The engineer and estimator determine the productivity rate. The rate is used for future cost analysis and estimating similar scopes of work.

Validate and clarify man-hour table using graphic analysis

The historical data collected in the field taken over time can be discrete or continuous. The collected data are summarized into a tabular arrangement by intervals that compacts the data into viewable records called a histogram. Plotting of data using graphic techniques allows results to be displayed in pictorial form. This provides insight into a data set to help with testing assumptions, model selection, regression model validation, estimator selection, relationship identification, factor effect determination, and outlier detection. Plotting of data allows the estimator to check assumptions in statistical models and communicate the results of an analysis.

The graph is a pictorial illustration of the relationship between variables:

$$Y = a + bx; \quad Y = a + (y - y1)/(x - x1)(x)$$

where

Y = dependent variable
a = intercept value along the y-axis at $x = 0$
b = slope, or the length of the rise divided by the length of the run
$b = (y - y1)/(x - x1)$
x = independent or control variable

Illustrative example of graphic analysis of data for boiler field tube welding

Foreman's Report

Project: Boiler Erection-Field Tube Welding		
Foreman: John Smith		Date: June 00, 0000
Craft: Boilermaker		Tube Welding—MH per Weld
Cost code	Phase code description	Design pressure (PSI)
	Tube size (OD)	1501–2000
000000	1.5	3.00
000000	2	3.70
000000	2.5	4.20
000000	3	4.80
000000	3.5	5.30
000000	4	5.90
000000	4.5	6.50
000000	5.5	7.70
000000	6.5	9.00

Graphic analysis of data using Excel's chart capabilities to plot the graph; to use Excel chart capabilities, highlight the range E98-F106 and **select Insert, Charts, Insert Scatter Chart, Chart Elements: Axis, Axis Titles, Chart Title, Gridlines, Legend, Trendline, More Options: Display equation on chart, Display R-squared value on chart** (Fig. 1).

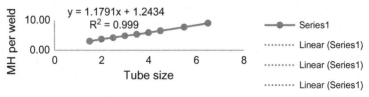

FIG. 1 Boiler field-type welding.

The coefficient of determination, R^2, is exactly +1 and indicates a positive fit.

All data points lie exactly on the straight line.

The relationship between X and Y variables is such that as X increases, Y also increases.

Boiler tube welding		
Field tube welding		
Facility—boiler		
	Tube welding—MH per weld	
x	*y*	
Tube size (OD)	*Design pressure (PSI)*	*Bins*
	1501–2000	
Over 1″ and including 1-1/2″ heli arc	3.00	0.11
Over 1-1/2″ and including 2″ heli arc	3.70	0.12
Over 2″ and including 2-1/2″ heli arc	4.20	0.30
Over 2-1/2″ and including 3″ heli arc	4.80	0.49
Over 3″ and including 3-1/2″ heli arc	5.30	0.60
Over 3-1/2″ and including 4″ heli arc	5.90	
Over 4″ and including 4-1/2″ heli arc	6.50	
Over 4-1/2″ and including 5-1/2″ heli arc	7.70	
Over 5-1/2″ and including 6-1/2″ heli arc	9.00	

Histogram

Use Excel's chart capabilities to plot the histogram (Fig. 2).

To use Excel chart capabilities, highlight range D156-E161, and select Insert, Charts.

Insert Scatter, Insert Combo Chart, Create Custom Combo Chart, Cluster Column.

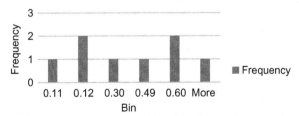

FIG. 2 Histogram.

Chart Elements: Axes, Axis Titles, Chart Title, Gridlines, Data Table.

Frequency distribution

Bin	Frequency
0.11	1
0.12	2
0.30	1
0.49	1
0.60	2
More	1

Regression analysis and correlation

Method of least squares

Linear regression: Fitting a straight line

The straight-line relationship can be valuable in summarizing the observed dependence of one variable on another. The most common type of linear regression is called ordinary least-squares regression. Linear regression uses the values from an existing data set consisting of measurements of the values of two variables, x and y, to develop a model that is useful for predicting the value of the dependent variable, y for given values of x. Thus, the statistical techniques used in linear regression can be used to uncover the relationships that exist in construction (Table 1).

Elements of a regression equation (linear, first-order model)

Regression equation: $Y = a + bx + \xi$

y is the value of the dependent variable (y), what is being predicted or explained.

a is a constant, equals the value of y when the value $x = 0$.

b is the coefficient of x, the slope of the regression line, how much y changes for each change in x.

ξ is the error term, the error in predicting the value of y, given the value of x.

Assumptions of linear regression

(1) Both the independent (x) and the dependent (y) variables are measured at interval or ration level.
(2) The relationship between the independent (x) and the dependent (y) variables is linear.
(3) Errors in the prediction of the value of y are distributed in a way that approaches the normal curve.
(4) Errors in the prediction of the value of y are all independent to one another.
(5) The distribution of the errors in the prediction of the value of y is constant regardless of the value of x.

Method of least squares

Least squares line

The least-squares line approximating the set of points (x1, y1), (x2, y2) ... (xn, yn) has the equation

$$Y = bx + a$$

where b = the slope of the line and a = y-intercept.

The best fit line for the $(x_1, y_1), (x_2, y_2) \dots (x_n, y_n)$ is given by

$$y - ybar = b(x - xbar)$$

where the slope is

$$b = \sum (xi - xbar)(yi - ybar) / \sum (xi - xbar)^2$$

and the y-intercept is

$$a = ybar - bxbar$$

Correlation

Correlation coefficient

(1) Positive correlation: If x and y have a strong positive linear correlation, r is close to +1. An r value exactly +1 indicates a perfect positive fit. Positive values indicate a relationship between x and y variables such that as values for x increase, values for y also increase.
(2) A perfect correlation of + or −1 occurs only when all the data points all lie exactly on a straight line. If r = +1, the slope line is positive. If r = −1, the slope line is negative.
(3) A correlation greater than 0.8 is generally described as strong, whereas a correlation less than 0.5 is generally described as weak.

Coefficient of determination, r^2 or R^2

(1) It is useful because it gives the proportion of the variance of one variable that is predictable from the other variable. It is a measure that allows us to determine how certain one can be of making predications from a certain model or graph.
(2) The coefficient of determination is the ratio of the explained variation to the total variation.
(3) The coefficient of determination is such that $(0 <= r^2 <= 1)$ and denotes the strength of the linear association between x and y.
(4) The coefficient of determination is a measure of how well the regression line represents the data. If the regression line passes exactly through every point on the scatter plot, it explains all of the variation. The further the line is away from the points, the less it is able to explain the variation.

Illustrative example of regression analysis and correlation of data for generating bank tube installation... Verify and clarify man-hour table

Foreman's Report			
Project: Waste Heat Boiler, Generating Bank Tube Installation			
Foreman: John Smith		*Date: June 00, 0000*	
Craft: Boilermaker			
Craft: Boilermaker Cost code **Generating bank tubes** 2-1/2″ OD × 0.203″ wall thickness, swage, and roll	Phase code description	QTY	MH
Install generating tubes	000000	2629	0.50
2′1/2″ Ends—expand tubes in steam and mud drums	000000	5258	0.44

Data for Input; Man-Hours for field erection for generating bank tube installation
Quantity (y): $R_1 = 2629, 5258$
Man-Hour (x): $R_2 = 0.50, 0.44$

TABLE 1 Linear regression: fitting a straight line

	MH/EA	Qty
Description	X	y
Generating bank tubes		
2-1/2″ OD × 0.203″ Wall thickness, swage, and roll		
Install generating tubes	0.50	2629
2′1/2″ Ends—expand tubes in steam and mud drums	0.44	5258
Covar (R_1 and R_2)	−39.44	
Varp (R_2)	0.00	
Slope (R_1 and R_2)	−43816.67	
Intercept (R_1 and R_2)	24537.33	
Correl (R_1 and R_2)=correlation coefficient	−1.0000	
Correl (R_1 and R_2) ^{2} = coefficient determination	1.0000	

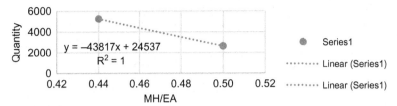

The coefficient of determination, R^2, is exactly $+1$ and indicates a positive fit. All data points lie exactly on the straight line. The relationship between x and y is such that as x increases, y also increases.

Nonrepetitive one-cycle time study and man-hour analysis

Nonrepetitive time study was used for construction direct craft long-cycle scopes of work. The time study provides man-hour information for cost estimating. The study requires continuous timing with an electric timer or video camera. A camcorder with video playback is used, and video tape can be returned to determine acceptable methods and the time required for the work. Once the time study is complete, the craft foreman determines the operation or task and calculates the time for the operation or task. The calculation is the net time less any reductions for task unrelated to the timed task. Normal time is found by multiplying a selected time for the task or cycle by the rating factor:

$$T_n = T_o \times RF$$

where

T_n = normal time, hours
T_o = observed time, hours
RF = rating factor, arbitrarily set, number

Example: If the craft worker is fast, then the $RF > 1.0$, say example as 110%. Task time is 1.8 h; then, normal time is $(1.8 \times 1.10) = 1.98$ h.

If the worker is rated 90%, then $RF < 1$ and normal time is $(0.90 \times 1.8) = 1.62$ h. The rating factor allows the "sample" observation to be adjusted for normal workers to arrive at a true value. Normal time does not include factors that affect labor productivity. Allowances for these factors are divided into three components: personal, fatigue, and delay (PF&D) process of timing the cycle: Idle time is excluded; craft takes breaks for coffee and rest room; allowance for personal is 5%. Fatigue is physiological reduction in ability to do work; allowance for fatigue is 5% delays beyond the worker's ability to prevent; allowance for delays is 5%. Productivity time in the work day is inversely proportional to the amount of PF&D allowance; the allowance is expressed as a percent of the total work day. PF&D allowance is generally in the range of 10%–20%.

Allowance multiplier:

$$F_a = 100\% / (100\% - PF\&D\%)$$

where $F_\ni$ = allowance multiplier for PF&D, number PF&D = personal, fatigue, and delay allowance, percentage standard productivity—time required by a trained and motivated worker or workers to perform construction task while working at normal tempo.

$$H_s = T_n \times F_\ni$$

where H_s = standard time for a construction task per unit of effort, hour. The allowance for PF&D is 15%, which is an allowance multiplier of 1.176.

Definitions for nonrepetitive one-cycle time study are given in Table 2.

TABLE 2 Definitions for productivity analysis

Task/operation	Designated and described work sublet to work measurement, estimating, and reporting.
Normal time	An element or operation time found by multiplying the average time observed for one or multiple cycles by a rating factor.
Element	A subpart of an operation or task separated for timing and analysis; beginning and ending points are described, and the element is the smallest part of an operation observed by time study. The length of the element can vary from minutes, to hours, to days, etc.
Rating factor	A means of comparing the performance of the worker under observation by using experience or other benchmarks; additionally, a numerical factor is noted for the elements or cycle; 100% is normal, and rating factors less than or greater than normal indicate slower or faster performance.
Observed time	The time observed on the stopwatch/electronic clock or other media and recorded on the time study sheet or media tape during the measurement process.
Cycle	The total time of elements from start to finish.
Continuous time	A method of time study where the total elapsed time from the start is recorded at the of each element.
Standard time	Sums of rated elements that have been increased for allowances.
Allowance	An adjustment for work, which includes delays, idle time, and inefficiency, as well as efficient effort.
Delay allowance	One part of the allowance included in the standard time for interruptions or delays beyond which the worker's ability to prevent.
Fatigue allowance	An allowance based on physiological reduction in ability to do work, sometimes included in the standard time.
Idle time	An interval in which the worker, equipment, or both are not performing useful work.
Person/man-hour	A unit of measure representing one person working for one hour.

TABLE 2 Definitions for productivity analysis—cont'd

Productivity	The amount of work performed in a given period. Usually measured in units of work per man- hour, man-month, or man-year.
Actual time	The time reported for work, which includes delays, idle time, and inefficiency, as well as efficient effort.
Productivity rate	A unit rate of production; the total amount produced in a given period divided by the number of hours, months, or years.

Code of accounts; tracking system for diesel power plant, fire water system (Tables 3 and 4)

TABLE 3 Illustration of portion of job cost by cost code and type for erection of fire water system

		Budget	
Cost code	Phase code description	Craft	MH
Fire water system (A14) estimate			
000000	Fire hose equipment	IW	130
000000	Hose cabinet	IW	30
000000	Fire water hydrant	IW	260
000000	Extinguisher	IW	80
000000	Mobile foam unit	IW	40
	Direct craft man-hours		**540**

TABLE 4 Illustration of portion of tracking report for erection of a fire water system

Cost code	Phase code description	Budget	Actual	Percent complete	Productivity (A/E)
Fire water system (A14) estimate					
000000	Fire hose equipment	130	125	100.00%	1.04
000000	Hose cabinet	30	31	100.00%	0.97

Continued

TABLE 4 Illustration of portion of tracking report for erection of a fire water system—cont'd

Cost code	Phase code description	Budget	Actual	Percent complete	Productivity (A/E)
000000	Fire water hydrant	260	257	100.00%	1.01
000000	Extinguisher	80	82	100.00%	0.98
000000	Mobile foam unit	40	40	100.00%	1.00
	Direct craft man-hours	**540**	**535**	**100.00%**	1.01

Estimating methods

Comparison method

The comparison logic is based on estimating similarities and differences for proposed equipment installation and previously installed equipment for which historical man-hour data are available.

Designate

MHc = Historical man-hour (know man-hour estimate)
MH_a = Equipment quantity is increased
MH_β = Equipment quantities are decreased

MHc $<=$ MH_a, MH_a is directly proportional to MHc; MH_β is a lower bound, and logic is expanded to $MH_\beta <=$ MHc $<= MH_a$.

The estimate that is either above or below the known estimate must satisfy the work scope and quantity takeoff. The less difference between the equipment estimate and the original comparison estimate, the better the comparison.

The comparison method provides the opportunity to compare the proposed estimate with the previous estimate. The estimate data must be current and need to be verified by measurement, project cost reports, one-cycle repetitive time study, and historical experience, and the data need to be revised continuously with the many combinations of construction design and engineering, labor skill, material, equipment, and procedures. The comparison method in this manual will be more effective if maintained and revised by the contractor, engineer, and owner.

Estimating equipment by comparison

Determine scope of work for the proposed equipment installation and make a detailed quantity takeoff.

The quantity takeoff is either above or below the known quantity and satisfies the scope.

The lower or upper bound is the reference to increased or decreased quantities. The known quantity is historical data from previous field installations.

The main advantages of comparison method

1. Estimate is based on comparison with actual unit man-hours and quantities.
2. The scope and quantity differences can be identified and the impacts estimated.
3. Comparison method applies to any complexity of design, bid, and contract for a project.

A simple example:

Estimator is using comparison to estimate duct work to be erected in a coal-fired power plant.

Coal-Fired Power Plant; Ductwork

$$MH_\beta <= MHc <= MH_\partial$$

where

MHc = historical man-hour
MH_∂ = equipment quantity is increased
MH_β = equipment quantities are decreased (Tables 5–7)

TABLE 5 MHc = historical man-hour estimate for over fire duct work

Estimate—duct work						Estimate
			Historical			
Description	MH	Qty	Unit	Qty	Unit	BM
Over fire air duct						<u>1930</u>
Transition duct section	32.00	20.0	Ton	20.0	Ton	640
Bolted connection-duct transition to ID fan	0.26	200.0	EA	200.0	EA	52
Duct sections	32.00	8.0	Ton	8.0	Ton	256
Duct elbow	32.00	8.0	Ton	8.0	Ton	256
Field joint-weld flange	0.35	8.0	LF	8.0	LF	3
Bolted connection-duct sections	0.26	12.0	EA	12.0	EA	3
Expansion joint	40.00	12.0	Ton	12.0	Ton	480
Duct support	40.00	6.0	Ton	6.0	Ton	240

TABLE 6 MH_a = estimate for increased over fire duct work quantities

Estimate—duct work						Estimate
			Historical			
Description	MH	Qty	Unit	Qty	Unit	BM
Over fire air duct						2100
Transition duct section	32.00	25.0	Ton	25.0	Ton	800
Bolted connection-duct transition to ID fan	0.26	240.0	EA	240.0	EA	62
Duct sections	32.00	9.0	Ton	8.0	Ton	256
Duct elbow	32.00	9.0	Ton	8.0	Ton	256
Field joint-weld flange	0.35	9.0	LF	8.0	LF	3
Bolted connection-duct sections	0.26	14.0	EA	12.0	EA	3
Expansion joint	40.00	12.0	Ton	12.0	Ton	480
Duct support	40.00	7.0	Ton	6.0	Ton	240

TABLE 7 MH_β = estimate for decreased over fire duct work quantities

Estimate—duct work						Estimate
			Historical			
Description	MH	Qty	Unit	Qty	Unit	BM
Over fire air duct						1712
Transition duct section	32.00	25.0	Ton	18.0	Ton	576
Bolted connection-duct transition to ID fan	0.26	190.0	EA	190.0	EA	49
Duct sections	32.00	7.0	Ton	7.0	Ton	224
Duct elbow	32.00	8.0	Ton	8.0	Ton	256
Field joint-weld flange	0.35	9.0	LF	9.0	LF	3
Bolted connection-duct sections	0.26	13.0	EA	13.0	EA	3
Expansion joint	40.00	10.0	Ton	10.0	Ton	400
Duct support	40.00	5.0	Ton	5.0	Ton	200

Comparison of graphic analysis of data (Fig. 3)

Comparison of estimated direct craft man-hours	Man-hours
MH_β = equipment quantities are decreased; estimated man-hours are decreased	1712
MHc = know unit man-hour based on historical data	1930
MH_a = equipment quantity is increase; estimated man-hours are increased	2100

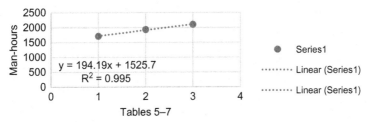

FIG. 3 Comparison of direct craft man-hours.

MHc is based on historical data from field erection of over fire duct work in a coal-fired power plant, and the scope and quantity differences can be identified and the impacts estimate; therefore,

$$MH_\beta <= MHc <= MH_a$$

The proposed unit is based on the estimator's quantity takeoff, and erection quantities are either ($\pm$) 10%.

Direct proportion (straight-line graph): comparison method quantities are directly proportional to estimate MHs.

MHc is directly proportional to MH_β and MH_a.

Whenever one variable increases or decreases, the other increases or decreases and vice versa.

Equation of straight line:

y is directly proportional to quantity x if y = kx for some k > 0.

Therefore, quantities are directly proportional to estimate MHs.

Unit quantity method

Many estimating methods are qualitative and depend on few facts. The unit method uses historical and quantitative evidence that leads to a cost driver easily understood. This method is important for detailed estimating for assemblies.

The method starts with the quantity takeoff arranged in the erection sequence required to assemble and install the equipment. The estimator selects the task description by defining the work scope for the task to be installed. Each task is related to and performed by direct craft and divided into one or more subsystems that are further divided into assemblies made up of construction line items.

The unit-quantity model is given by

$$MH_t = \sum n_i \, (MH_i)$$

where

MH_t = man-hours for equipment and piping field installation
n_i = task takeoff quantity i, in dimensional units
MH_i = unit equipment/piping man-hour associated with n_i
i = Task 1, 2, m from quantity takeoff associated w/ field equipment/piping

The n_i quantity is the takeoff for the field scope of work.

MH_i unit man-hours are determined from piping and major equipment man-hour tables.

The unit man-hours were determined from historical data, and they correspond to the labor productivity necessary to install piping and equipment in an industrial construction facility.

The unit-quantity method allows a final cross-check of actual man-hours to estimated man-hours.

Factor:

$$MH_t = n_i \, (MH_i)(f_u)$$

where f_u = factor percent for productivity loss and alloy weld factor; u = 1, 2, ..., p (Table 8).

TABLE 8 Illustration of unit-quantity method

Coal-fired power plant and over fire duct work installation

Estimate—over fire duct work

Scope of work—quantity takeoff	Quantity n_i		Man-hour MH_i	Total MH MH_t
Over fire air duct				
Transition duct section	20.0	Ton	32.0	640.0
Bolted connection-duct transition to ID fan	200.00	EA	0.26	52.0
Duct sections	8.0	Ton	32.0	256.0
Duct elbow	8.0	Ton	32.0	256.0
Field Joint-weld flange	8.0	LF	0.4	2.8
Bolted connection-duct sections	12.00	EA	0.26	3.1
Expansion joint	12.0	Ton	40.0	480.0
Duct support	6.0	Ton	40.0	240.0
Man-hour total				1930

Factors affecting labor productivity and impacts of overtime

The American Association of Cost Engineers defines productivity as a "relative measure of labor efficiency, either good or bad, when compared with an established base or norm." Estimates are based on a defined work scope, duration, start date and clarifications, and exceptions to the bid documents. The design may be incomplete, or changes are made that will impact the bid estimate.

Examples of impacts that affect labor productivity are as follows:

Project has added work scope and owner request project completed on bid date. Request may require increased craft manpower, a second shift, overtime, and many other impacts to the schedule and estimate.

The increased man-hours, extended schedule, and other resources will impact the schedule and project cost.

Delay of owner provided material and equipment to be install will affect the erection sequence, duration, and schedule of work packages. Reassignment of craft is required.

These impacts will cause an increase in manpower and joint occupancy with other trades causing a drop in productivity.

Labor factors

The estimator can use the following table of labor factors and values for factoring labor productivity for work packages.

The overtime labor factors provide a basis for estimating inefficiencies utilizing quantitative data actually measured on construction projects.

The principles and techniques in this manual will enable the experienced estimator to apply the methods to measure, analyze, and compile accurate estimates for industrial projects (Tables 9–11).

TABLE 9 Project overtime labor factors

Week of extended OT	50 h/week	60 h/week	72 h/week	84 h/week
1	0.05	0.09	0.14	0.25
2	0.07	0.12	0.2	0.30
3	0.08	0.14	0.27	0.35
4	0.09	0.19	0.32	0.40
5	0.15	0.24	0.37	0.45
6	0.16	0.28	0.42	0.50
7	0.24	0.33	0.46	0.53
8	0.23	0.36	0.49	0.56
9	0.26	0.38	0.50	0.57
10	0.28	0.39	0.51	0.58
11	0.28	0.40	0.52	0.59
12	0.29	0.41	0.53	0.60
13	0.31	0.44	0.54	0.61
14	0.32	0.45	0.55	0.62
15	0.33	0.46	0.56	0.63
16	0.34	0.47	0.57	0.64

TABLE 10 Illustration for project labor factors for productivity loss

	Crew size			Factor		Percent of base MH
Total craft man-hours						
Crew size man-hour factor	25					
Subtotal				1.00%	17180 171.8 17351.8 =	101.00%

Overtime factors

Week	Hours of exposure	Crew size MH>40	Impact MH	Factor	Additional hours
1	50	10	500	0.05	25
2	50	10	500	0.07	35
3	50	10	500	0.08	40
4	50	10	500	0.09	45
5	50	10	500	0.15	75
6	50	10	500	0.16	80
7					
8					
9					
10					
Subtotal hours of exposure	300				
Subtotal—additional overtime man-hours					300

TABLE 11 Straight time and overtime additional MHs

Other factors		MH	Factor			
Straight time additional MHs	MHs	17180	10%		1718	
Overtime additional MHs	MHs	600	5%		30	Percent of
					19400	MH
					2220	112.92%

	Range of impact			Use		
Other factors	*Minor*	*Average*	*Severe*	*ST*	*OT*	
Stacking of trades	10%	20%	30%	10%		
Morale and attitude	5%	15%	30%			
Reassignment of manpower	5%	10%	15%			
Stacking of contractors own forces	10%	20%	30%			
Dilution of supervision	10%	15%	25%			
Learning curve	5%	15%	30%			
Errors and omissions	3%	5%	6%			
Sharing occupancy with owners operations	5%	15%	25%			
Joint occupancy with other trades	5%	12%	20%		5%	
Interferences with access to work site	5%	12%	30%			
Deficient materials delivery	10%	25%	35%			
Fatigue	8%	10%	12%			

TABLE 11 Straight time and overtime additional MHs—cont'd

| Other factors | Range of impact | | | Use | |
	Minor	Average	Severe	ST	OT
Ripple	8%	20%	30%		
Season/ weather	10%	20%	30%		
Availability of skilled labor					
Ratio of name calls to open calls					
Plant permit requirements					
Safety					
				<u>10%</u>	<u>5%</u>

Work sampling and man hour analysis

Work sampling is a statistical technique for determining the proportion of time spent by craft crews performing manual construction task. In a work sampling study, a large number of observations are made of workers over an extended period of time. The observations must be taken at random times during the period of study. Work sampling is the application of statistical sampling techniques to the study of construction labor activities, measurement of labor efficiency, and productivity information.

Application of work sampling and man-hour analysis
Model:

$$Hs = (N_i)\,(Ht)\,(RF)\,(1 + PF\&D)/N$$

where

Hs = standard man-hours per task
N_i = observation of event i
Ht = total man-hours worked during sample study
RF = rating factor
PF&D = personal, fatigue, and delay allowance
N = number of random observations during sample study

Construction task: Boilermakers are connecting boiler tubes in the mud and steam drums.

The work is expected to take several months, and the work sampling occurs at the start of the overall work.

The study runs for 3 weeks for a crew of 14 boilermakers.

Productivity is determined to be $RF=0.95$. A total of $N=100$ tubes are installed and $N_i=0.61$ for an allowance of 20% per tube.

Calculate input data:

$$N_i = 0.61$$
$$Ht = 3 \times 14 \times 40$$
$$RF = 0.95$$
$$PF\&D = 1.2$$
$$N = 100$$
$$Hs = (0.61) * (3 * 14 * 40) * (0.95) * (1.2)/100$$
$$Hs = 11.7 \text{ man-hours per tube}$$

Chapter 1

Introduction to construction statistics using Excel

1.1 Section introduction—Introduction to construction statistics using Excel

This section provides individuals in construction and engineering a basic understanding of how to set up and use excel templates and spreadsheets. Excel is a powerful spreadsheet program that is used to perform mathematical and statistical calculations. The use of computer spreadsheets—specifically Microsoft Excel—demonstrates how templates and spreadsheets can be used to automate statistical functions. In addition, this section will provide an introduction to the basic operation of Excel. This includes how to set up Excel templates and spreadsheets, formatting spreadsheets, writing statistical and mathematical formulas, using the basic statistical functions, entering data, and printing results and graphs. The purpose of the following Excel statistical templates and spreadsheets is to make the connection to construction forecasting, learning curves, work sampling, risk analysis, and bidding strategy. To get the most benefits from the statistical applications, the reader should understand exponents, logarithms, and simple algebraic manipulations. It also helps, but is not required, to understand regression analysis. Readers whose mathematics is limited can plug the statistical functions into Excel templates and spreadsheet to get the results they need. The use of mathematics and statistics in construction is necessary to reduce risk in the bidding process and validate historical data to develop detailed accurate estimates. The applications should be of considerable value as a reference for those engaged in design, estimating, forecasting, and bidding projects in the industrial construction industry.

Industrial Process Plant Construction Estimating and Man-Hour Analysis.
https://doi.org/10.1016/B978-0-12-818648-0.00001-6

1.2 Graphic analysis of data

A graph is a pictorial illustration of the relationship between variables. Assume for historical man-hour data that x is the independent variable and y is the dependent variable:

1.2.1 Straight-line graph

$$y = a + bx; \quad Y = a + (y - y1)/(x - x1)(x)$$

where

y = dependent variable
a = intercept value along the y-axis at $x = 0$
b = slope, or the length of the rise divided by the length of the run; $b = (y - y1)/(x - x1)$
x = independent or control variable

Example 1.2.1

Create a straight-line graph for the historical data (Table 1.2.1).

TABLE 1.2.1 0.375″ or less, carbon steel BW, SMAW—downhill

Facility—diesel power plant	MH/JT
x	y
Pipe size	MH/BW
0.5	0.80
0.75	0.80
1	0.80
1.5	0.80
2	0.80
2.5	0.88
3	1.05
4	1.40
6	2.10
8	2.80
10	3.50
12	4.20
14	4.90
16	5.60
18	6.30
20	7.00
24	8.40

Use Excel's chart capabilities to plot the graphical straight line given by the equation y = a + bx.

To use the Excel chart capabilities, highlight the range C44-D60, and select **Insert, Scatter Chart, Chart Elements, Axis, Axis Titles, Chart Title, Gridlines, Legend, Trendline, and More Options: Display equation on chart and Display R-squared value on chart** (Fig. 1.2.1).

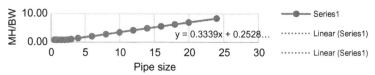

FIG. 1.2.1   CS BW-downhill.

The coefficient of determination, R^2, is exactly +1 and indicates a positive fit, and all points lie exactly on the straight line. The relationship between X and Y variables is such that as X increases and Y increases.

1.3   Linear regression and method of least squares

1.3.1   Linear regression: Fitting a straight line

The straight-line relationship can be valuable in summarizing the observed dependence of one variable on another. The most common type of linear regression is called ordinary least-squares regression. Linear regression uses the values from an existing data set consisting of measurements of the values of two variables, x and y, to develop a model that is useful for predicting the value of the dependent variable, y for given values of x. Thus, the statistical techniques used in linear regression can be used to uncover the relationships that exist in construction.

1.3.2   Elements of a regression equations (linear, first-order model)

Regression equation:

$$y = a + bx + \varepsilon$$

y is the value of the dependent variable (y), what is being predicted or explained.
a, a constant, equals the value of y when the value of $x = 0$.
b is the coefficient of X, the slope of the regression line, how much Y changes for each change in x.
x is the value of the independent variable (x), what is predicting or explaining the value of y.
ε is the error term, the error in predicting the value of y, given the value of x

1.3.3   Assumptions of linear regression

1. Both the independent (x) and the dependent (y) variables are measured at interval or ration level.
2. The relationship between the independent (x) and the dependent (y) variables is linear.
3. Errors in prediction of the value of y are distributed in a way that approaches the normal curve.
4. Errors in prediction of the value of y are all independent of one another.
5. The distribution of the errors in prediction of the value of y is constant regardless of the value of x.

1.3.4   Method of least squares

Least square line

The least-squares line approximating the set of points (x1, y1), (x2, y2) ... (xn, yn) has the equation

$$y = bx + a$$

where

b = the slope of the line
a = y-intercept

The best fit line for the points $(x1, y1), (x2, y2) \ldots (xn, yn)$ is given by

$$y - \bar{y} = b(x - \bar{x})$$

where the slope is

$$b = \Sigma (xi - \bar{x})(yi - \bar{y}) / \Sigma (xi - \bar{x})^2$$

and the y-intercept is

$$a = \bar{y} - b\bar{x}$$

1.4 Using Excel to forecast boiler tube replacement

Excel statistical functions for forecasting the value of y for any x. Thus, a and b can be calculated in Excel, where

R1 = the array of y values and R2 = the array of x values

b = SLOPE (R1,R2) = COVAR (R1,R2)/VARP (R2)

a = INTERCEPT (R1,R2) = AVERAGE (R1)−b * Average (R2)

SLOPE (R1,R2) = slope of regression line

INTERCEPT (R1,R2) = y-intercept of the regression line

FORECAST (x, R1,R2) calculates the predicted value of y for given value of x. Thus, FORECAST (x, R1,R2) = a +b * x where a = INTERCEPT (R1, R2) and b = SLOPE (R1, R2).

TREND (R1,R2) = array function that produces an array of predicted y values corresponding to x values stored in array R2, based on the regression line calculated from x values stored in array R2 and y values stored in array R1.

COVAR (R1,R2) = returns covariance, the average of the products of deviations for each data point pair in two data cells

VARP (R2) = calculates variance based on the entire population (ignores logical values and text in the population

Using Excel statistical functions set up Table 1.4.1 **and** Fig. 1.4.1

To use the Excel statistical functions, go to quick access toolbar, select **More Functions,** and then, select **Statistical**.

Excel statistical functions are used to calculate values for data. **Set up Template 1.4.1.**

To use **COVAR (R1, R2)**, highlight the array of values for x (array R1), enter a comma, and highlight the array of values y (array R2).

To use **VARP (R2)**, highlight the array of values for x (array R2).

To use **SLOPE (R1, R2)**, highlight the array of values for y (array R1), enter a comma, and highlight the array of values x (array R2).

To use **INTERCEPT (R1, R2)**, highlight the array of values for y (array R1), enter a comma, and highlight the array of values x (array R2).

Template 1.4

Calculate the regression line for the historical data in Example 1.4 and plot the results.

Scope of field work for boiler tube replacement
    Remove and install new boiler generating tubes
    Boiler scaffold, rigging, drum internals and remove boiler trim
    Remove soot blowers and casing
    Remove tubes and install new tubes
    Reinstall casing, trim, soot blowers, test, and clean up.
Data for input: man-hours for field installation of new boiler generating tubes
    Quantity (y): R1 = 1.0, 2.0, 4.0, and 5.0
    Man-hour (x): R2 = 172, 210, 525.2, and 746

TABLE 1.4.1 Linear regression: Fitting a straight line man-hour quantity

Description	x	y
Boiler tube replacement		
Boiler scaffold, rigging, drum internals, and remove boiler trim	172	1.0
Remove soot blowers and casing	210	2.0
Remove tubes and install new tubes	515.2	4.0
Reinstall casing, trim, soot blowers, test and clean up	746	5.0

COVAR (R1, R2) = COVAR (D172-D175; E172-E175)   363.30
VARP (R2) = VARP (E172-E175)   55151.1200
SLOPE (R1, R2) = Slope (E172-E175; D172-D175)   0.0066
INTERCEPT (R1, R2) = INTERCEPT (E172-E175; D172-D175)   0.2939
Excel chart capabilities are used to plot the regression line. **Set** Fig. 1.4.1.
Use Excel's chart capabilities to plot the graphical straight line given by the equation $y = a + bx$.

To use the Excel chart capabilities, highlight the range D172-E175, and select **Insert, Scatter Chart, Chart Elements, Axis, Axis Titles, Chart Title,**

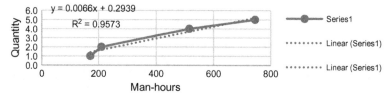

FIG. 1.4.1   Install new boiler generating tubes.

Gridlines, Legend, Trendline, and More Options: Display equation on chart and Display R-squared value on chart.

CORREL (R1, R2) = CORREL (D172-D175; E172-E175)	0.9784
CORREL (R1, R2) ˆ2 = CORREL (D172-D175; E172-E175) ˆ2	0.9573

The coefficient of determination is $R^2 = 0.9573$, and the correlation coefficient, $R = -0.9784$, is a strong indication of correlation. 97.8% of the total variation on Y can be explained by the linear relationship between X and Y (described by the regression equation, $Y = 0.0066X + 0.2939$). The relationship between X and Y variables is such as X increases and Y increases.

1.5   Correlation

1.5.1   Correlation coefficient

(1) Measures the strength and direction of a linear relationship between two variables.

(2) Mathematical formula for computing r is

$$r = n\left(\sum XY\right) - \left(\sum X\right)\left(\sum Y\right) / \left[\left(n\left(\sum X^2 - \left(\sum X\right)^2\right)\right]\left[n\left(\sum Y^2 - \left(\sum Y\right)^2\right)\right]\right)^{1/2}.$$

(3) The value of r is such that $(-1 <= r <= +1)$. The $+$ and $-$ signs are used for positive and negative linear correlations, respectively.

(4) Positive correlation: If x and y have a strong positive linear correlation, r is close to +1. An r valve exactly +1 indicates a perfect positive fit. Positive values indicate a relationship between X and Y variables such that as values for X increase, values for Y also increase.

(5) A perfect correlation of $+$ or -1 occurs only when all the data points all lie exactly on a straight line. If $r = +1$, the slope line is positive. If $r = -1$, the slope line is negative.

(6) A correlation **greater than 0.8** is generally described as strong, whereas a correlation **less than 0.5** is generally described as weak.

1.5.2   Coefficient of determination, r^2

(1) It is useful because it gives the proportion of variance of one variable that is predictable from the other variable. It is a measure that allows us to determine how certain one can be of making predictions from a certain model/graph.

(2) The coefficient of determination is the ratio of the explained variation to the total variation.

(3) The coefficient of determination is such that $(o < = r^2 < = 1)$ and denotes the strength of the linear association between X and Y.

(4) The coefficient of determination represents the percent of the data that is the closest to the line of best fit. For example, if $r = 0.922$, then $r^2 = 0.850$, which means that 85% of the total variation in Y can be explained by the linear relationship between X and Y (as described by the regression equation). The other 15% of the total variation in Y remains unexplained.

(5) The coefficient of determination is a measure of how well the regression line represents the data. If the regression line passes exactly through every point on the scatter plot, it would explain all of the variation. The further the line is away from the points, the less it is able to explain.

1.5.3   Excel functions

Excel provides the following functions for the correlation coefficient and coefficient determination:

CORREL (R1, R2) = correlation coefficient of data in arrays R1 and R2
CORREL (R1, R2) $\hat{}$2 = coefficient determination

1.6 Time series

Time series is a set of observations taken at specified times, usually at equal intervals, and is defined by the values Y1, Y2, ... of a variable Y at times t1, t2, ... Thus, Y is a function of t, $Y = F(t)$. The analysis of time series consists of a mathematical description of component movements. Analysis of movements is of great value in many connections to construction, one of which is the problem of forecasting future movements. Thus, the statistical techniques used in forecasting time series can be used to uncover the relationships that exist in construction.

1.6.1 Simple moving average

Given a set of numbers

$$Y_1, Y_2, \dots$$

define a moving average of order m to be given by the sequence of arithmetic means:

$$Y1 + Y2 + \dots + My/m, Y2 + Y3 + \dots + ym/m, Y3 + Y4 + \dots + my/m, \dots$$

The sums in the numerators are moving totals of order m.

Data given annually or monthly, a moving average of order m is called respectively an m year average or m month average.

Moving averages have the property to reduce the amount of variation present in the data. In the case of time series, this property is used to eliminate fluctuations, and the process is called smoothing of time series.

Using a simple moving average model, forecast the next value(s) in a time series based on the average of a fixed finite number m of the previous values. Thus, for all $i > m$,

$$\hat{y}i = 1/m \, \Sigma \, yi = (yi - m + \dots + yi - 1)/m$$

1.6.2 Excel functions

AVERAGE (number 1, number 2): returns the average (arithmetic mean) of its arguments, which can be numbers or names, arrays, or references that contain numbers.

ABS (number): returns the absolute value of a number, a number without its sign.

Using Excel statistical functions set up Table 1.6.1

To use the Excel statistical functions, go to quick access toolbar, select **more functions,** and then, select **statistical.**

Excel statistical functions are used to calculate values for data. **Set up Template 1.**

To produce the values, insert the formulas =**AVERAGE** (D297-D299), =**ABS** (D300-E300), =(D300-E300) 2 in cells F619, G619, and H619, respectively. Insert the =AVERAGE (F299-F04) for **MAE** (cell F307) and =**AVERAGE** (G299-G304) for **MSE** (cell G307); then, highlight the range E300-FG04 and press **Ctrl-D.**

Example 1.6.1

Calculate the forecast values for the time series using a simple moving average with m = 3

Data for input: man-hours for the field erection of a waste heat boiler

Year (t): 1005, 2006, 2007, 2008, 2009, 2010, 2011, and 2012

Average man-hour (y): 23351, 24864, 23929, 22670, 24515, 26360, 25053, and 24992

TABLE 1.6.1 Simple moving average forecast

Year		Average		
		Man-hours		
t	y	Predict	(e)	e̅2
2005	23351			
2006	24864			
2007	23929			
2008	22670	24047.9	1377.9	1898565.0
2009	24515	23299.3	1215.6	1477770.9
2010	26360	23592.5	2767.3	7658135.0
2011	25053	25437.4	384.7	147975.0
2012	24992	25706.2	713.8	509446.7
			MAE	MSE
			1291.9	2338378.5

Excel chart capabilities are used to plot the simple moving average. **Set** Fig. 1.6.1.

Use Excel's chart capabilities to plot the graph for the simple moving average forecast.

To use the Excel chart capabilities, highlight the range D300-E307, and select **Insert,** and select from **Recommended Charts, all charts, line, chart elements, and axis title.**

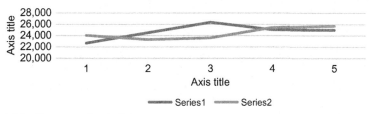

FIG. 1.6.1 Erect waste heat boiler.

The plot of forecast values [predict in *red (light gray)*] smoothed out the plot of y values [in *blue (dark gray)*]. The higher the value of m, the more smoothing that occurs.

1.6.3 Simple exponential smoothing

A statistical technique for working out averages while allowing for recent changes in values by moving the period under consideration at regular intervals. In simple exponential smoothing, the forecast value for Y at time $i + 1$ that is made at time t equals the simple average of the most recent m observations. In particular, for some α where $0 <= \alpha <= 1$ for all $i > 1$, define

$$\hat{y}_1 = y_1 \quad \hat{y}_i + 1 = \hat{y}_i + ae_i$$

Iteration is expressed as

$$\hat{y}_1 = y_1 \quad \hat{y}i + 1 = ay_i + (1 - a)\hat{y}_i$$

Using a simple exponential smoothing, forecast the next value(s) with $\alpha = 0.2$.

Most recent period's demand multiplied by the smoothing factor, plus the most recent period's forecast multiplied by (one minus the smoothing factor)

Formula:

$$F_t = \alpha^*D + (1 - \alpha)^*F$$

where

$\quad$ D = most recent period's demand
$\quad$ α = smoothing factor in decimal form
$\quad$ F = most recent period's forecast

Enter formula in cell E371 is =D371, and the formula in cell E372 is =D382*D371+(1−D382)*E371 copy formula in cell E373 to cell E378.

Enter formula in cell F372 is = D372-E372, and the formula in cell G372 is =F372 2; copy formula in cell F373 to cell F378 and formula in cell G372 to cell G378.

Example 1.6.2

Using Example 1.3 (simple moving average); using exponential smoothing with $\alpha = 0.2$ (Table 1.6.2)

Data for input: man-hours for the field erection of waste heat boiler

Year (t): 2005, 2006, 2007, 2008, 2009, 2010, 2011, and 2012

Average man-hour (y): 23351, 24864, 23929, 22670, 24515, 26360, 25053, and 24992

TABLE 1.6.2 Simple exponential smoothing forecast

Year	Man-hour			
t	y	Predict	e	e^2
2005	23351	23351		
2006	24864	23351	1513.2	2289771.4
2007	23929	23654	274.9	75565.9
2008	22670	23709	−1038.6	1078673.6
2009	24515	23501	1014.0	1028227.1
2010	26360	23704	2656.1	7054874.2
2011	25053	24235	817.8	668733.6
2012	24992	24398	594.0	352852.2
	Predict	189902.3		
	Alpha		MAE	MSE
	0.2		833.1	1792671.1

Excel chart capabilities are used to plot the simple exponential smoothing. **Set** Fig. 1.6.2.

Use Excel's chart capabilities to plot the graph for the simple exponential smoothing.

To use the Excel chart capabilities, highlight the range D365-E372, and select **Insert,** and select from **Recommended Charts, all charts, line, chart elements, and axis title.**

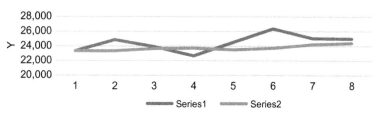

FIG. 1.6.2 Exponential smoothing field erection waste heat boiler.

The plot of forecast values [predict in *red (light gray)*] smoothed out the plot of y values [in *blue (dark gray)*]. The lower the value of α, the more smoothing that occurs.

1.7 The mean, variance and standard deviation measures of central tendency statistical formulas for the mean, variance and standard deviation

Mean: The arithmetic mean of a set of data is denoted by $\overline{Y}$; it is simply the arithmetic average of all the observations; the formula for the mean is

$$\overline{Y} = (Y1 + Y2 + \cdots + Yn)/n$$

$\overline{y}$ can be written

$$\overline{y} = \sum_{i=1}^{n} yi/n$$

Excel function:
Excel function: The mean is calculated in Excel using the function **AVERAGE**.
AVERAGE (number 1, number 2): Returns the average (arithmetic mean) of its arguments, which can be numbers or names, arrays, or references that contain numbers.

Using Excel statistical functions set up Table 1.7.1

To use the Excel statistical functions, go to quick access toolbar; select **More Functions,** and then, select **Statistical**.
Excel statistical functions are used to calculate values for data. **Set up Template 1.7.1**
To produce the values, insert the formulas =**AVERAGE** (C423-C429).
Calculate the mean ($\overline{y}$), CS schedule 40 BW man-hour.

Data for input:
$\overline{y}$: 1.10, 1.10, 2.02, 2.99, 4.07, 5.15, and 6.02
n: 7

TABLE 1.7.1 Mean ($\overline{y}$) CS schedule 40 BW MH

Mean	
$\overline{y}$	
1.10	
1.10	
2.02	
2.99	
4.07	
5.15	
6.02	
AVERAGE (C423-C429)=	3.21

1.7.1 The mean, variance and standard deviation measures of central tendency

The formula for the variance is

$$S^2 = \left(y1 - \overline{Y}\right)^2 + \left(Y2 + \overline{Y^2}\right) + \cdots + \left(Yn - \overline{Y}\right)^2 / n - 1$$

and summation notation

$$S^{\wedge}2 = \sum_{i=1}^{n} (yi - \overline{y})^{\wedge}2 / n - 1$$

Excel function: The sample variance is calculated in Excel using the function **VAR**.

VAR (number 1, number 2): Estimates variance based on a sample (ignores logical values and text in the sample).

Using Excel statistical functions set up Table 1.7.2

To use the Excel statistical functions, go to quick access toolbar; select **more functions,** and then, select **statistical**. Excel statistical functions are used to calculate values for data. **Set up Template 1.7.2.**

To produce the values, insert the formulas =**VAR** (C566-C572).

Calculate the variance, CS schedule 40 BW man-hour.

Data for input: s^2: 1.10, 1.10, 2.02, 2.99, 4.07, 5.15, and 6.02.

TABLE 1.7.2 Variance CS schedule 40 BW MH	
Mean	
$\overline{y}$	
1.10	
1.10	
2.02	
2.99	
4.07	
5.15	
6.02	
VAR (C463-C469)=	3.79

1.7.2 The mean, variance and standard deviation measures of central tendency

The **standard deviation** is a measure of how spread out the numbers are and is the square root of the variance. The formula for the standard deviation is

$$S = \left[(y1 - \overline{Y})^2 + Y2 + \overline{Y}^2 \right) + \cdots + (Yn - \overline{Y})^2 / n - 1 \right] \frac{1}{2}$$

$$s = \left[\sum_{i=1}^{n} (yi - \overline{y})\text{^}2 / n - 1 \right] \text{^}1/2$$

Excel function:
Excel function: The standard deviation is calculated in Excel using the function **STDEV**.
STDEV (number 1, number 2): Estimates standard deviation based on a sample (ignores logical values and text in the sample)

To use the Excel statistical functions, go to quick access toolbar; select **More Functions,** and then, select **Statistical**.

To produce the values, insert the formulas **=STDEV** (C463-C469) and **STDEV (C463-C469) = 1.95.**

Chapter 2

Process piping

2.1 Section introduction—Piping schedules and tables

This section contains schedules and tables that cover the complete labor for the field fabrication and installation of process piping in an industrial facility. The direct craft man-hours for handling, welding, and bolting pipe are based on welding methods, pipe wall thickness, pressure, and temperature. The piping schedules for large bore pipe units for handling pipe are in diameter inch-feet, and the units for welding, bolting, and handling valves are in diameter inches. Piping man-hour tables, for carbon steel piping, are derived from the schedules. The actual man-hour production units were developed from job cost reports for process piping installed in the field on construction projects. The projects involved large crew sizes, a low ratio of regular employees, and a safe workplace that strives for zero accidents. Each of these factors has a negative effect on productivity and has been considered in the development of the piping man-hours. Standardized man-hours are required so that all work is from the same "baseline" data. The baseline can be modified, and the standard man-hours can be set up in a computerized estimating system. Man-hours are for direct craft labor and do not include craft supervision and indirect craft support man-hours. Refer to Schedule D—alloy and nonferrous weld factors to apply the percentages for field fabrication and erection of alloy and nonferrous piping.

2.2 Piping section general notes

Handling and erecting pipe—man-hours to unload, store in lay down, haul, rig, and align in place.

Field handling valves—man-hours to place screwed, flanged, and weld end valves and expansion joints.

Field handling control valves and specialty items—man-hours are two times manual valves per valve.

Field erection bolt ups—man-hours per joint to bolt-up valves, expansion joints, flanged fittings, and spools.

Making on screwed fittings and valves—man-hours for cutting, threading, handling, and erection per connection.

Industrial Process Plant Construction Estimating and Man-Hour Analysis.
https://doi.org/10.1016/B978-0-12-818648-0.00002-8

2.2.1 General welding notes

Manual butt welds—wall thickness of the pipe determines man-hours that will apply per joint.

Postweld heat treatment (PWHT) butt welds greater than or equal to 0.750″.
PWHT craft support—man-hours for craft to warp/and remove pipe insulation for stress relief.
Apply percentages for welding alloys and nonferrous butt welds.
Preheating and stress relieving butt welds are not included.

Olet welds—man-hours are two times butt weld and include cutting, placing, and welding per connection.

Stub in weld—man-hours are 1.5 times butt weld and include cutting, placing, and welding per connection.

Socket welds—man-hours include fit up and welding per joint.

Hydrotest pipe—man-hours to place/remove blinds, open/close valves, removal/replacement of valves and specialty items and pipe sections as required and drain lines after testing.

2.3   Schedule A—Oil refinery piping

Standard labor estimating units

Facility—oil refinery Description	Large bore piping Unit of measure Man-hours per unit	Small bore piping Unit of measure Man-hours per unit
Handle and install pipe, carbon steel, **welded joint**	Diameter inch-feet	MH/LF
WT <= 0.375″	0.06	0.15
0.406″ >= WT <= 0.500″	0.07	0.18
0.562″ >= WT <= 0.688″	0.08	0.20
0.718″ <= WT <= 0.938″	0.11	0.28
1.031″ <= WT <= 1.219″	0.20	0.50
1.250″ <= WT <= 1.312″	0.25	0.75
Welding, carbon steel, arc-uphill	Diameter inch	MH/EA
WT <= 0.375″	0.45	1.00
0.406″ >= WT <= 0.500″	0.50	1.15
0.562″ >= WT <= 0.688″	0.95	1.90
0.718″ >= WT <= 0.938″ (PWHT)	1.10	2.25
1.031″ >= WT <= 1.218″ (PWHT)	1.35	2.60
1.250″ <= WT <= 1.312″ (PWHT)	2.10	4.20
SOL, TOL, WOL	2 × BW	2 × BW
Stub in	1.5 × BW	1.5 × BW
Socket weld		Per SW table
PWHT craft support labor	0.45	1.00
Bolt up of flanged joints by weight class	Diameter inch	MH/EA
150#/300# Bolt up	0.40	1.00
600#/900# Bolt up	0.50	1.20
1500# Bolt up	0.65	1.60
Handle valves by weight class	Diameter inch	MH/EA
150# and 300# manual valve	0.50	1.00
600# and 900# manual valve	0.52	1.30
Heavier manual valve >= 1500#	1.00	2.00
Control valve and specialty items	2 × manual	2 × manual

2.3.1 Handle and install pipe, carbon steel, welded joint

Facility—Oil refinery

Man-hours per foot

Pipe size	Wall thickness in inches				
	Up to 0.375″	0.406″–0.500″	0.562″–0.688″	0.718″–0.938″	1.031″–1.219″
0.5	0.15	0.18	0.20	0.28	0.50
0.75	0.15	0.18	0.20	0.28	0.50
1	0.15	0.18	0.20	0.28	0.50
1.5	0.15	0.18	0.20	0.28	0.50
2	0.15	0.18	0.20	0.28	0.50
2.5	0.15	0.18	0.20	0.28	0.50
3	0.18	0.21	0.24	0.33	0.60
4	0.24	0.28	0.32	0.44	0.80
6	0.36	0.42	0.48	0.66	1.20
8	0.48	0.56	0.64	0.88	1.60
10	0.60	0.70	0.80	1.10	2.00
12	0.72	0.84	0.96	1.32	2.40
14	0.84	0.98	1.12	1.54	2.80
16	0.96	1.12	1.28	1.76	3.20
18	1.08	1.26	1.44	1.98	3.60
20	1.20	1.40	1.60	2.20	4.00
24	1.44	1.68	1.92	2.64	4.80

Erect pipe man-hours include all labor to unload, store in lay down, haul, rig, and align in place.

2.3.2   Welding butt weld, carbon steel, and SMAW—Uphill

Facility—Oil refinery

Man-hours per joint

Pipe size	Wall thickness in inches				
	Up to 0.375″	0.406″–0.500″	0.562″–0.688″	0.718″–0.938″	1.031″–1.219″
0.5	1.00	1.15	1.90	2.25	2.60
0.75	1.00	1.15	1.90	2.25	2.60
1	1.00	1.15	1.90	2.25	2.60
1.5	1.00	1.15	1.90	2.25	2.60
2	1.00	1.15	1.90	2.25	2.60
2.5	1.13	1.25	2.38	2.75	3.38
3	1.35	1.50	2.85	3.30	4.05
4	1.80	2.00	3.80	4.40	5.40
6	2.70	3.00	5.70	6.60	**8.10**
8	3.60	4.00	7.60	*8.80*	*10.80*
10	4.50	5.00	9.50	**11.00**	**13.50**
12	5.40	6.00	11.40	**13.20**	**16.20**
14	6.30	7.00	13.30	**15.40**	**18.90**
16	7.20	8.00	15.20	**17.60**	**21.60**
18	8.10	9.00	17.10	**19.80**	**24.30**
20	9.00	10.00	19.00	**22.00**	**27.00**
24	10.80	12.00	22.80	**26.40**	**32.40**

Manual butt welds.
Wall thickness of the pipe determines man-hours that will apply per joint.
PWHT butt welds greater than or equal to 0.750″.
Apply percentages for welding alloys and nonferrous butt welds.

2.3.3 Welding—SOL, TOL, and WOL

Facility—Oil refinery

Man-hours per SOL, TOL, and WOL

Pipe size	Wall thickness in inches				
	Up to 0.375″	0.406″–0.500″	0.562″–0.688″	0.718″–0.938″	1.031″–1.219″
0.5	2.00	2.30	3.80	4.50	5.20
0.75	2.00	2.30	3.80	4.50	5.20
1	2.00	2.30	3.80	4.50	5.20
1.5	2.00	2.30	3.80	4.50	5.20
2	2.00	2.30	3.80	4.50	5.20
2.5	2.25	2.50	4.75	5.50	6.75
3	2.70	3.00	5.70	6.60	8.10
4	3.60	4.00	7.60	8.80	10.80
6	5.40	6.00	11.40	13.20	**16.20**
8	7.20	8.00	15.20	**17.60**	**21.60**
10	9.00	10.00	19.00	**22.00**	**27.00**
12	10.80	12.00	22.80	**26.40**	**32.40**
14	12.60	14.00	26.60	**30.80**	**37.80**
16	14.40	16.00	30.40	**35.20**	**43.20**
18	16.20	18.00	34.20	**39.60**	**48.60**
20	18.00	20.00	38.00	**44.00**	**54.00**
24	21.60	24.00	45.60	**52.80**	**64.80**

SOL, TOL, and WOL welds; two times BW (includes cut, prep, and fit up).
Wall thickness of the pipe used for weld determines man-hours that will apply.
Apply percentages for welding alloys and nonferrous welds.

2.3.4   PWHT craft support labor

Facility—Oil refinery

Man-hour per joint

Pipe size	MH
0.5	1.00
0.75	1.00
1	1.00
1.5	1.00
2	1.00
2.5	1.13
3	1.35
4	1.80
6	2.70
8	3.60
10	4.50
12	5.40
14	6.30
16	7.20
18	8.10
20	9.00
24	10.80

Place and remove insulation from weld joint.

2.3.5 Field bolt up of flanged joints by weight class

Facility—Oil refinery

Man-hours per joint

	Pressure rating		
Pipe size	150#/300#	600#/900#	1500#/2500#
0.5	1.00	1.20	1.60
0.75	1.00	1.20	1.60
1	1.00	1.20	1.60
1.5	1.00	1.20	1.60
2	1.00	1.20	1.60
2.5	1.00	1.25	1.63
3	1.20	1.50	1.95
4	1.60	2.00	2.60
6	2.40	3.00	3.90
8	3.20	4.00	5.20
10	4.00	5.00	6.50
12	4.80	6.00	7.80
14	5.60	7.00	9.10
16	6.40	8.00	10.40
18	7.20	9.00	11.70
20	8.00	10.00	13.00
24	9.60	12.00	15.60

Man-hours per joint to bolt-up valves, expansion joints, flanged fittings, and spools.

2.3.6   Field handle valves/specialty items by weight class

Facility—Oil refinery

Pipe size	Man-hours per valve/specialty			Control valve/specialty		
	Manual valve pressure rating			Pressure rating		
	150#/300#	600#/900#	1500#/2500#	150#/300#	600#/900#	1500#/2500#
0.5	1.00	1.30	2.00	2.00	2.60	4.00
0.75	1.00	1.30	2.00	2.00	2.60	4.00
1	1.00	1.30	2.00	2.00	2.60	4.00
1.5	1.00	1.30	2.00	2.00	2.60	4.00
2	1.00	1.30	2.00	2.00	2.60	4.00
2.5	1.25	1.30	2.50	2.50	2.60	5.00
3	1.50	1.56	3.00	3.00	3.12	6.00
4	2.00	2.08	4.00	4.00	4.16	8.00
6	3.00	3.12	6.00	6.00	6.24	12.00
8	4.00	4.16	8.00	8.00	8.32	16.00
10	5.00	5.20	10.00	10.00	10.40	20.00
12	6.00	6.24	12.00	12.00	12.48	24.00
14	7.00	7.28	14.00	14.00	14.56	28.00
16	8.00	8.32	16.00	16.00	16.64	32.00
18	9.00	9.36	18.00	18.00	18.72	36.00
20	10.00	10.40	20.00	20.00	20.80	40.00
24	12.00	12.48	24.00	24.00	24.96	48.00

Man-hours only—field handle and erect screwed, flanged, weld end valves, and expansion joints.

2.4 Schedule B—Diesel power plant piping

Standard labor estimating units

Facility—diesel power plant Description	Large bore piping Unit of measure Man-hours per unit	Small bore piping Unit of measure Man-hours per unit
	Diameter inch-feet	MH/LF
Handle and install pipe, carbon steel, welded joint		
WT<=0.375″	0.05	0.12
0.406″>=WT<=0.500″	0.06	0.15
0.562″>=WT<=0.688″	0.07	0.18
0.718″<=WT<=0.938″	0.10	0.25
1.031″<=WT<=1.219″	0.12	0.30
1.250″<=WT<=1.312″	0.15	0.35
Welding, carbon steel, arc-downhill		
WT<=0.375″	0.35	0.80
0.406″>=WT<=0.500″	0.40	1.00
0.562″>=WT<=0.688″	0.85	1.90
0.718″>=WT<=0.938″ (PWHT)	1.15	2.40
1.031″>=WT<=1.218″ (PWHT)	1.40	2.90
1.250″<=WT<=1.312″ (PWHT)	1.50	3.00
SOL, TOL, WOL	2 × BW	0.00
Stub in	1.5 × BW	0.00
Socket weld	0.65	
PWHT craft support labor	0.45	1.00
Bolt up of flanged joints by weight class		
150#/300# Bolt up	0.45	0.90
600#/900# Bolt up	0.55	1.20
1500# Bolt up	0.65	1.60
Handle valves by weight class		
150# and 300# manual valve	0.35	0.80
600# and 900# manual valve	0.60	1.30
Heavier manual valve>=1500#	0.90	1.80

2.4.1   Handle and install pipe, carbon steel, and welded joint

Facility—Diesel power plant

Man-hours per foot

Pipe size	Wall thickness in inches				
	0.375" or less	0.406"–0.500"	0.562"–0.688"	0.718"–0.938"	1.031"–1.219"
0.5	0.12	0.15	0.18	0.25	0.30
0.75	0.12	0.15	0.18	0.25	0.30
1	0.12	0.15	0.18	0.25	0.30
1.5	0.12	0.15	0.18	0.25	0.30
2	0.12	0.15	0.18	0.25	0.30
2.5	0.13	0.15	0.18	0.25	0.30
3	0.15	0.18	0.21	0.30	0.36
4	0.20	0.24	0.28	0.40	0.48
6	0.30	0.36	0.42	0.60	0.72
8	0.40	0.48	0.56	0.80	0.96
10	0.50	0.60	0.70	1.00	1.20
12	0.60	0.72	0.84	1.20	1.44
14	0.70	0.84	0.98	1.40	1.68
16	0.80	0.96	1.12	1.60	1.92
18	0.90	1.08	1.26	1.80	2.16
20	1.00	1.20	1.40	2.00	2.40
24	1.20	1.44	1.68	2.40	2.88

Erect pipe man-hours include all labor to unload, store in lay down, haul, rig, and align in place.

2.4.2   Welding butt weld, carbon steel, and SMAW—Downhill

Facility—Diesel power plant

Man-hours weld joint

Pipe size	Wall thickness in inches				
	0.375" or less	0.406"–0.500"	0.562"–0.688"	0.718"–0.938"	1.031"–1.219"
0.5	0.80	1.00	1.90	2.40	2.90
0.75	0.80	1.00	1.90	2.40	2.90
1	0.80	1.00	1.90	2.40	2.90
1.5	0.80	1.00	1.90	2.40	2.90
2	0.80	1.00	1.90	2.40	2.90
2.5	0.88	1.00	2.13	2.88	3.50
3	1.05	1.05	1.20	3.45	3.45
4	1.40	1.40	1.60	3.40	4.60
6	2.10	2.10	2.40	5.10	6.90
8	2.80	2.80	3.20	*6.80*	*9.20*
10	3.50	3.50	4.00	**8.50**	**11.50**
12	4.20	4.20	4.80	**10.20**	**13.80**
14	4.90	4.90	5.60	**11.90**	**16.10**
16	5.60	5.60	6.40	**13.60**	**18.40**
18	6.30	6.30	7.20	**15.30**	**20.70**
20	7.00	7.00	8.00	**17.00**	**23.00**
24	8.40	8.40	9.60	**20.40**	**27.60**

Manual butt welds.
Wall thickness of the pipe determines man-hours that will apply per joint.
PWHT butt welds greater than or equal to 0.750".
Apply percentages for welding alloys and nonferrous butt welds.
Preheating and stress relieving butt welds are not included.

2.4.3   Welding—SOL, TOL, and WOL

Facility—Diesel power plant

Man-hours per SOL, TOL, and WOL

Pipe size	Wall thickness in inches				
	0.375″ or less	0.406″–0.500″	0.562″–0.688″	0.718″–0.938″	1.031″–1.219″
0.5	1.60	2.00	3.80	4.80	5.80
0.75	1.60	2.00	3.80	4.80	5.80
1	1.60	2.00	3.80	4.80	5.80
1.5	1.60	2.00	3.80	4.80	5.80
2	1.60	2.00	3.80	4.80	5.80
2.5	1.75	2.00	4.25	5.75	7.00
3	2.10	2.10	2.40	6.90	6.90
4	2.80	2.80	3.20	6.80	9.20
6	4.20	4.20	4.80	10.20	**13.80**
8	5.60	5.60	6.40	**13.60**	**18.40**
10	7.00	7.00	8.00	**17.00**	**23.00**
12	8.40	8.40	9.60	**20.40**	**27.60**
14	9.80	9.80	11.20	**23.80**	**32.20**
16	11.20	11.20	12.80	**27.20**	**36.80**
18	12.60	12.60	14.40	**30.60**	**41.40**
20	14.00	14.00	16.00	**34.00**	**46.00**
24	16.80	16.80	19.20	**40.80**	**55.20**

SOL, TOL, and WOL welds; two times BW (includes cut, prep, and fit up).
Wall thickness of the pipe used for weld determines man-hours that will apply.
PWHT butt welds greater than or equal to 0.750″.
Apply percentages for welding alloys and nonferrous welds.

2.4.4 PWHT craft support labor

Facility—Diesel power plant

Man-hour per joint

Pipe size	MH per joint
0.5	1.00
0.75	1.00
1	1.00
1.5	1.00
2	1.00
2.5	1.63
3	1.95
4	2.60
6	3.90
8	5.20
10	6.50
12	7.80
14	9.10
16	10.40
18	11.70
20	13.00
24	15.60

Insulate and remove insulation from weld joint.

2.4.5   Field bolt up of flanged joints by weight class

Facility—Diesel power plant

Man-hours per joint

Pipe size	Pressure rating		
	150#/300#	600#/900#	1500#/2500#
0.5	0.90	1.20	1.60
0.75	0.90	1.20	1.60
1	0.90	1.20	1.60
1.5	0.90	1.20	1.60
2	0.90	1.20	1.60
2.5	1.13	1.38	1.63
3	1.35	1.65	1.95
4	1.80	2.20	2.60
6	2.70	3.30	3.90
8	3.60	4.40	5.20
10	4.50	5.50	6.50
12	5.40	6.60	7.80
14	6.30	7.70	9.10
16	7.20	8.80	10.40
18	8.10	9.90	11.70
20	9.00	11.00	13.00
24	10.80	13.20	15.60

Man-hours per joint to bolt-up valves, expansion joints, flanged fittings, and spools.

2.4.6 Field handle valves/specialty items by weight class

Facility—Diesel power plant

Pipe size	Man-hours per valve			Control valve/specialty		
	Manual valve pressure rating			Pressure rating		
	150#/ 300#	600#/ 900#	1500#/ 2500#	150#/ 300#	600#/ 900#	1500#/ 2500#
0.5	0.80	1.30	1.80	1.60	2.60	3.60
0.75	0.80	1.30	1.80	1.60	2.60	3.60
1	0.80	1.30	1.80	1.60	2.60	3.60
1.5	0.80	1.30	1.80	1.60	2.60	3.60
2	0.80	1.30	1.80	1.60	2.60	3.60
2.5	0.88	1.50	2.25	1.75	3.00	3.00
3	1.05	1.80	2.70	2.10	3.60	3.60
4	1.40	2.40	3.60	2.80	4.80	4.80
6	2.10	3.60	5.40	4.20	7.20	7.20
8	2.80	4.80	7.20	5.60	9.60	9.60
10	3.50	6.00	9.00	7.00	12.00	12.00
12	4.20	7.20	10.80	8.40	14.40	14.40
14	4.90	8.40	12.60	9.80	16.80	16.80
16	5.60	9.60	14.40	11.20	19.20	19.20
18	6.30	10.80	16.20	12.60	21.60	21.60
20	7.00	12.00	18.00	14.00	24.00	24.00
24	8.40	14.40	21.60	16.80	28.80	28.80

Man-hours only—field handle and erect screwed, flanged, weld end valves, and expansion joints.

2.5   Schedule C—Solar power plant piping

Standard labor estimating units

Facility—solar power plant Description	Large bore piping Unit of measure Man-hours per unit	Small bore piping Unit of measure Man-hours per unit
Handle and install pipe, carbon steel, **welded joint**	**Diameter inch-feet**	**MH/LF**
WT<=0.375″	0.07	0.18
0.406″<=WT<=0.500″	0.09	0.23
0.562″<=WT<=0.688″	0.11	0.28
0.718″<=WT<=0.938″	0.14	0.35
1.031″<=WT<=1.219″	0.20	0.50
1.250″<=WT<=1.312″	0.25	0.75
Welding butt welds, carbon steel, **arc-uphill**	**Diameter inch**	**MH/EA**
WT<=0.375″	0.50	1.10
0.406″>=WT<=0.500″	0.55	1.20
0.562″>=WT<=0.688″	1.05	2.20
0.718″>=WT<=0.938″ (PWHT)	1.20	2.45
1.031″>=WT<=1.219″ (PWHT)	1.45	2.70
1.250″<=WT<=1.312″ (PWHT)	2.20	4.40
Olet- (SOL, TOL, and WOL)	2 × BW	2 × BW
Stub in	1.5 × BW	1.5 × BW
Socket weld		Per SW table
PWHT craft support labor	0.45	1.00
Bolt up of flanged joints by weight class	**Diameter inch**	**MH/EA**
150#/300# Bolt up	0.40	1.00
600#/900# Bolt up	0.50	1.20
1500#/2500# Bolt up	0.65	1.60
Handle valves by weight class	**Diameter inch**	**MH/EA**
150# and 300# manual valve	0.45	1.00
600# and 900# manual valve	0.90	1.80
Heavier manual valve>=1500#	1.80	2.00

2.5.1 Handle and install pipe, carbon steel, and welded joint

Facility—Solar power plant

Man-hours per foot

Pipe size	Wall thickness in inches				
	0.375″ or less	0.406″–0.500″	0.562″–0.688″	0.718″–0.938″	1.031″–1.219″
0.5	0.18	0.23	0.28	0.35	0.50
0.75	0.18	0.23	0.28	0.35	0.50
1	0.18	0.23	0.28	0.35	0.50
1.5	0.18	0.23	0.28	0.35	0.50
2	0.18	0.23	0.28	0.35	0.50
2.5	0.18	0.23	0.28	0.35	0.50
3	0.21	0.27	0.33	0.42	0.60
4	0.28	0.36	0.44	0.56	0.80
6	0.42	0.54	0.66	0.84	1.20
8	0.56	0.72	0.88	1.12	1.60
10	0.70	0.90	1.10	1.40	2.00
12	0.84	1.08	1.32	1.68	2.40
14	0.98	1.26	1.54	1.96	2.80
16	1.12	1.44	1.76	2.24	3.20
18	1.26	1.62	1.98	2.52	3.60
20	1.40	1.80	2.20	2.80	4.00
24	1.68	2.16	2.64	3.36	4.80

Erect pipe man-hours include all labor to unload, store in lay down, haul, rig, and align in place.

2.5.2   Welding butt weld, carbon steel, and SMAW—Uphill

Facility—Solar power plant

Man-hour per joint

Pipe size	Wall thickness in inches				
	0.375″ or less	0.406″–0.500″	0.562″–0.688″	0.718″–0.938″	1.031″–1.219″
0.50	1.10	1.20	2.20	2.45	2.70
0.75	1.10	1.20	2.20	2.45	2.70
1.00	1.10	1.20	2.20	2.45	2.70
1.50	1.10	1.20	2.20	2.45	2.70
2.00	1.10	1.20	2.20	2.45	2.70
2.50	1.25	1.38	2.63	3.00	3.63
3.00	1.50	1.65	3.15	3.60	4.35
4.00	2.00	2.20	4.20	4.80	5.80
6.00	3.00	3.30	6.30	7.20	8.70
8.00	4.00	4.40	8.40	9.60	11.60
10.00	5.00	5.50	10.50	12.00	14.50
12.00	6.00	6.60	12.60	14.40	17.40
14.00	7.00	7.70	14.70	16.80	20.30
16.00	8.00	8.80	16.80	19.20	23.20
18.00	9.00	9.90	18.90	21.60	26.10
20.00	10.00	11.00	21.00	24.00	29.00
24.00	12.00	13.20	25.20	28.80	34.80

Manual butt welds.
Wall thickness of the pipe determines man-hours that will apply per joint.
PWHT butt welds greater than or equal to 0.750′.
Apply percentages for welding alloys and nonferrous butt welds.

2.5.3 Welding—SOL, TOL, and WOL

Facility—Solar power plant

Man-hour Per SOL, TOL, and WOL

Pipe size	Wall thickness in inches				
	0.375″ or less	0.406″–0.500″	0.562″–0.688″	0.718″–0.938″	1.031″–1.219″
0.5	2.20	2.40	4.40	4.90	5.40
0.75	2.20	2.40	4.40	4.90	5.40
1	2.20	2.40	4.40	4.90	5.40
1.5	2.20	2.40	4.40	4.90	5.40
2	2.20	2.40	4.40	4.90	5.40
2.5	2.50	2.75	5.25	6.00	7.25
3	3.00	3.30	6.30	7.20	8.70
4	4.00	4.40	8.40	9.60	11.60
6	6.00	6.60	12.60	14.40	**17.40**
8	8.00	8.80	16.80	**19.20**	**23.20**
10	10.00	11.00	21.00	**24.00**	**29.00**
12	12.00	13.20	25.20	**28.80**	**34.80**
14	14.00	15.40	29.40	**33.60**	**40.60**
16	16.00	17.60	33.60	**38.40**	**46.40**
18	18.00	19.80	37.80	**43.20**	**52.20**
20	20.00	22.00	42.00	**48.00**	**58.00**
24	24.00	26.40	50.40	**57.60**	**69.60**

SOL, TOL, and WOL welds; two times BW (includes cut, prep, and fit up).
Wall thickness of the pipe used for olet determines man-hours that will apply.
Apply percentages for welding alloys and nonferrous welds.

2.5.4   Welding stub in

Facility—Solar power plant

Man-hour per stub in

Pipe size	Wall thickness in inches				
	0.375" or less	0.406"–0.500"	0.562"–0.688"	0.718"–0.938"	1.031"–1.219"
0.5	1.65	1.80	3.30	3.68	4.05
0.75	1.65	1.80	3.30	3.68	4.05
1	1.65	1.80	3.30	3.68	4.05
1.5	1.65	1.80	3.30	3.68	4.05
2	1.65	1.80	3.30	3.68	4.05
2.5	1.88	2.06	3.94	4.50	5.44
3	2.25	2.48	4.73	5.40	6.53
4	3.00	3.30	6.30	7.20	8.70
6	4.50	4.95	9.45	10.80	**13.05**
8	6.00	6.60	12.60	**14.40**	**17.40**
10	7.50	8.25	15.75	**18.00**	**21.75**
12	9.00	9.90	18.90	**21.60**	**26.10**
14	10.50	11.55	22.05	**25.20**	**30.45**
16	12.00	13.20	25.20	**28.80**	**34.80**
18	13.50	14.85	28.35	**32.40**	**39.15**
20	15.00	16.50	31.50	**36.00**	**43.50**
24	18.00	19.80	37.80	**43.20**	**52.20**

Stub in—1.5 times BW (includes cut, prep, and fit up).
Wall thickness of the pipe used for stub in determines man-hours that will apply.
Apply percentages for welding alloys and nonferrous welds.

2.5.5 PWHT craft support labor

Facility—Solar power plant

Man-hour per joint

Pipe size	MH
0.5	1.00
0.75	1.00
1	1.00
1.5	1.00
2	1.00
2.5	1.13
3	1.35
4	1.80
6	2.70
8	3.60
10	4.50
12	5.40
14	6.30
16	7.20
18	8.10
20	9.00
24	10.80

Place and remove insulation from weld joint.

2.5.6   Field bolt up of flanged joints by weight class

Facility—Solar power plant

Man-hour per joint

Pipe size	Pressure rating		
	150#/300#	600#/900#	1500#/2500#
0.5	1.00	1.20	1.60
0.75	1.00	1.20	1.60
1	1.00	1.20	1.60
1.5	1.00	1.20	1.60
2	1.00	1.20	1.60
2.5	1.00	1.25	1.63
3	1.20	1.50	1.95
4	1.60	2.00	2.60
6	2.40	3.00	3.90
8	3.20	4.00	5.20
10	4.00	5.00	6.50
12	4.80	6.00	7.80
14	5.60	7.00	9.10
16	6.40	8.00	10.40
18	7.20	9.00	11.70
20	8.00	10.00	13.00
24	9.60	12.00	15.60

Man-hours per joint to bolt-up valves, expansion joints, flanged fittings, and spools.

2.5.7 Field handle valves/specialty items by weight class

Facility—Solar power plant

Pipe size	Man-hours per valve/specialty			Control valve/specialty		
	Pressure rating			Pressure rating		
	150#/300#	600#/900#	1500#/2500#	150#/300#	600#/900#	1500#/2500#
0.5	1.00	1.80	2.00	2.00	3.60	4.00
0.75	1.00	1.80	2.00	2.00	3.60	4.00
1	1.00	1.80	2.00	2.00	3.60	4.00
1.5	1.00	1.80	2.00	2.00	3.60	4.00
2	1.00	1.80	2.00	2.00	3.60	4.00
2.5	1.13	2.25	4.50	2.25	4.50	9.00
3	1.35	2.70	5.40	2.70	5.40	10.80
4	1.80	3.60	7.20	3.60	7.20	14.40
6	2.70	5.40	10.80	5.40	10.80	21.60
8	3.60	7.20	14.40	7.20	14.40	28.80
10	4.50	9.00	18.00	9.00	18.00	36.00
12	5.40	10.80	21.60	10.80	21.60	43.20
14	6.30	12.60	25.20	12.60	25.20	50.40
16	7.20	14.40	28.80	14.40	28.80	57.60
18	8.10	16.20	32.40	16.20	32.40	64.80
20	9.00	18.00	36.00	18.00	36.00	72.00
24	10.80	21.60	43.20	21.60	43.20	86.40

Field handle and erect screwed, flanged, weld end valves, and expansion joints.

2.6   Schedule D—Alloy and nonferrous weld factors

Welding percentages for alloy and nonferrous metals

Pipe size	Material classification-group numbers							
	1	2	3	4	5	6	7	8
2	0.25	0.54	0.20	0.58	2.11	2.25	0.225	0.45
3	0.275	0.58	0.23	0.61	2.15	2.32	0.25	0.495
4	0.30	0.61	0.25	0.68	2.22	2.35	0.28	0.54
5	0.315	0.63						0.57
6	0.345	0.65	0.30	0.75	2.28	2.40	0.30	0.62
8	0.39	0.74	0.50	0.88	2.38	2.50	0.34	0.70
10	0.425	0.85	0.75	0.95	2.45	2.75	0.375	0.765
12	0.45	2.00	0.80	2.04	2.50	3.00	0.40	0.81
14	0.49	2.15						0.88
16	0.525	2.23						0.945
18	0.59	2.30						2.06
20	0.65	2.45						2.17
24	0.73							2.24

Group 1—Chrome molybdenum steel, chrome—1/2%–13%, moly—TO 1%
Group 2—18-8 Stainless steel, TY, 304, 316, and 347
Group 3—Copper, brass, and everdur
Group 4—Aluminum, monel, and copper, chrome-nickel

2.7   Schedule E—Industrial plant piping

Standard labor estimating units

Facility—industrial plant Description	Large bore piping Unit of measure Man-hours per unit	Small bore piping Unit of measure Man-hours per unit
	Diameter inch	MH/EA
PVC pipe		
Handle and install PVC pipe	0.046	0.12
PVC solvent joints	0.15	0.35
Copper pipe		
Handle and install copper pipe	0.10	0.18
Copper sweat joints	0.70	1.40
Victaulic pipe		
Victaulic couplings	0.35	0.70
Victaulic grooving	0.35	0.70
Mechanical couplings (dresser)	0.50	1.00
Screwed pipe		
Handle and install screwed pipe	0.035	
Screwed joints (measure, cut, thread, and makeup)	0.30	
Miscellaneous		
Hydrotest	$0.12 \times$ pipe handle	
In-line instruments (PI's and TI's)		1.20
Socket weld		Per SW table

2.7.1   Welding, carbon steel, and socket weld

Facility—Industrial plant

Man-hour per SW

Pipe size (inches)	Socket welds	
	Schedule 40 and 80	Schedule 100 and greater
0.5	1.2	1.3
0.75	1.2	1.3
1	1.3	1.4
1.5	1.4	1.6
2	1.5	1.8

Includes place and fit up.

2.7.2   Hydrostatic testing

Facility—Industrial plant

Man-hours per lineal foot

Pipe size	Wall thickness in inches				
	0.375″ or less	0.406″–0.500″	0.562″–0.688″	0.718″–0.938″	1.031″–1.219″
0.5	0.022	0.028	0.034	0.042	0.060
0.75	0.022	0.028	0.034	0.042	0.060
1	0.022	0.028	0.034	0.042	0.060
1.5	0.022	0.028	0.034	0.042	0.060
2	0.022	0.028	0.034	0.042	0.060
2.5	0.021	0.027	0.033	0.042	0.060
3	0.025	0.032	0.040	0.050	0.072
4	0.034	0.043	0.053	0.067	0.096
6	0.050	0.065	0.079	0.101	0.144
8	0.067	0.086	0.106	0.134	0.192
10	0.084	0.108	0.132	0.168	0.240
12	0.101	0.130	0.158	0.202	0.288
14	0.118	0.151	0.185	0.235	0.336
16	0.134	0.173	0.211	0.269	0.384
18	0.151	0.194	0.238	0.302	0.432
20	0.168	0.216	0.264	0.336	0.480
24	0.202	0.259	0.317	0.403	0.576

Man-hours to place/remove blinds, open/close valves, removal/replacement of valves and specialty items and pipe sections as required, and drain lines after testing.

2.7.3   Cut and prep weld joint

Facility—Industrial plant

Man-hours per joint

Pipe size	Wall thickness in inches				
	0.375″ or less	0.406″–0.500″	0.562″–0.688″	0.718″–0.938″	1.031″–1.219″
0.5	0.00	0.00	0.00	0.00	0.00
0.75	0.00	0.00	0.00	0.00	0.00
1	0.00	0.00	0.00	0.00	0.00
1.5	0.00	0.00	0.00	0.00	0.00
2	0.00	0.00	0.00	0.00	0.00
2.5	0.38	0.45	0.50	0.58	0.68
3	0.45	0.54	0.60	0.69	0.81
4	0.60	0.72	0.80	0.92	1.08
6	0.90	1.08	1.20	1.38	1.62
8	1.20	1.44	1.60	1.84	2.16
10	1.50	1.80	2.00	2.30	2.70
12	1.80	2.16	2.40	2.76	3.24
14	2.10	2.52	2.80	3.22	3.78
16	2.40	2.88	3.20	3.68	4.32
18	2.70	3.24	3.60	4.14	4.86
20	3.00	3.60	4.00	4.60	5.40
24	3.60	4.32	4.80	5.52	6.48

2.7.4   PVC piping

Facility—Industrial plant

Pipe size	Handle pipe	Solvent joint	Valve handling		Bolt up MH/joint
	MH/LF	MH/joint	150#/300#	C.V./SP.	150#/300#
0.5	0.12	0.35	0.40	0.80	1.00
0.75	0.12	0.35	0.40	0.80	1.00
1	0.12	0.35	0.40	0.80	1.00
1.5	0.12	0.35	0.40	0.80	1.00
2	0.12	0.35	0.40	0.80	1.00
2.5	1.88	0.38	0.63	1.25	1.00
3	2.25	0.45	0.75	1.50	1.20
4	3.00	0.60	1.00	2.00	1.60
6	4.50	0.90	1.50	3.00	2.40
8	6.00	1.20	2.00	4.00	3.20
10	7.50	1.50	2.50	5.00	4.00
12	9.00	1.80	3.00	6.00	4.80

Erect pipe man-hours include all labor to unload, store in lay down, haul, rig, and align in place.
Solvent joints include cut, square, ream, fit up, and make joint.
Handle valves—man-hours include field handling and erection of valves.

2.7.5   Copper piping

Facility—Industrial plant

	Handle pipe	Sweat joint					
Pipe size	MH/LF	MH/ joint	150#/ 300#	600#/ 900#	C.V./ SP.	150#/ 300#	600#/ 900#
0.5	0.18	1.40	0.45	0.90	1.80	1.00	1.30
0.75	0.18	1.40	0.45	0.90	1.80	1.00	1.30
1	0.18	1.40	0.45	0.90	1.80	1.00	1.30
1.5	0.18	1.40	0.75	1.50	3.00	1.00	1.30
2	0.18	1.40	0.90	1.80	3.60	1.00	1.30
2.5	0.25	1.75	1.13	2.25	4.50	1.00	1.50
3	0.30	2.10	1.35	2.70	5.40	1.20	1.80
4	0.40	2.80	1.80	3.60	7.20	1.60	2.40
6	0.60	4.20	2.70	5.40	10.80	2.40	3.60
8	0.80	5.60	3.60	7.20	14.40	3.20	4.80
10	1.00	7.00	4.50	9.00	18.00	4.00	6.00
12	1.20	8.40	5.40	10.80	21.60	4.80	7.20

Erect pipe man-hours include all labor to unload, store in lay down, haul, rig, and align in place.
Man-hours include handling and jointing or making on solder-type brass or copper joints.
Handle valves—man-hours include field handling and erection of valves.
Man-hours per joint to bolt-up valves.

2.7.6   Screwed pipe

Facility—Industrial plant

Pipe size	Handle pipe	Screwed joint
Inches	MH/LF	MH/Jt
0.5	0.10	0.30
0.75	0.10	0.30
1	0.10	0.30
1.5	0.10	0.45
2	0.10	0.60

Handle and install screwed pipe.
Screwed joints (measure, cut, thread, and makeup).

2.8   Schedule K—Underground drainage piping for industrial plants

Standard labor estimating units

Facility—industrial plant (underground drainage piping) Description	Unit of measure Man-hours per unit
Cast iron—lead and mechanical joint	**Diameter inch-feet**
Handle and install pipe	0.03
Handle and install fittings	3′ × Fitting
	MH/dia in
Lead and mechanical joints	0.20
	Diameter inch-feet
Cast iron soil—no hub	
Handle and install pipe	0.025
Handle and install fittings	3′ × Fitting
	MH/dia in
Couplings	0.10
Pipe trench	
	MH/CY
Trench excavation—backhoe 1 CY bucket	0.15
Trench excavation—backhoe 3/4 CY bucket	0.20
Trench excavation—backhoe 1/2 CY bucket	0.25
Trench excavation—hand (to 4′ deep)	2.40
Backfill and compaction—loader and compactors	0.60
Backfill and compaction—hand and compactors	3.40
Sand bedding and shading with loader	0.40
Shoring (place/remove)	MH/SF
Laborer	5.50
Carpenter	3.00
	Diameter inch-feet
Vitrified clay pipe	0.02
Make-ons	0.08
Concrete pipe	0.018
Cement poured Jt	0.06

2.8.1   Handle and install pipe and cast iron—Lead and mechanical joint

Facility—Industrial plant (underground drainage piping)

Pipe size	Pipe set and align	Lead and mechanical joint	Handle install fitting
Inches	MH/LF	MH/JT	MH/EA
4	0.12	0.80	0.24
6	0.18	1.20	0.36
8	0.24	1.60	0.48
10	0.30	2.00	0.60
12	0.36	2.40	0.72
12	0.36	2.40	0.72
16	0.48	3.20	0.96
18	0.54	3.60	1.08
20	0.60	4.00	1.20
24	0.72	4.80	1.44

Pipe man-hours include handle, haul, place, and align in trench.

2.8.2   Handle and install pipe and cast iron soil—No hub

Facility—Industrial plant (underground drainage piping)

Pipe size	Pipe set and align	Handle install fitting	Couplings
Inches	MH/LF	MH/EA	MH/EA
4	0.10	0.24	0.40
6	0.15	0.36	0.60
8	0.20	0.48	0.80
10	0.25	0.60	1.00
12	0.30	0.72	1.20
12	0.30	0.72	1.20
16	0.40	0.96	1.60
18	0.45	1.08	1.80
20	0.50	1.20	2.00
24	0.60	1.44	2.40

Pipe man-hours include handle, haul, place, and align in trench.

2.8.3   Handle and install pipe, concrete, and vitrified clay

2.8.3.1   Facility—Industrial plant (underground drainage piping)

Pipe size	Vitrified clay pipe		Concrete pipe	
	Pipe set and align	Make-on	Pipe set align	Cement poured joint
Inches	MH/LF	MH/JT	MH/LF	MH/JT
4	0.08	0.32	0.07	0.24
6	0.12	0.48	0.11	0.36
8	0.16	0.64	0.14	0.48
10	0.20	0.80	0.18	0.60
12	0.24	0.96	0.22	0.72
12	0.24	0.96	0.22	0.72
16	0.32	1.28	0.29	0.96
18	0.36	1.44	0.32	1.08
20	0.40	1.60	0.36	1.20
24	0.48	1.92	0.43	1.44
30	0.60	2.40	0.54	1.80
36	0.72	2.88	0.65	2.16
42	0.84	3.36	0.76	2.52
48	0.96	3.84	0.86	2.88
60	1.20	4.80	1.08	3.60

Pipe man-hours include handle, haul, place, and align in trench.

2.8.4   Pipe trench—Machine, hand excavation, backfill, and sand bedding

Pipe trench—machine and hand excavation	
Facility—industrial plant (underground drainage piping)	
Machine excavation	MH/CY
Trench excavation—backhoe 1 CY bucket	0.15
Trench excavation—backhoe 3/4 CY bucket	0.20
Trench excavation—backhoe 1/2 CY bucket	0.25
Hand excavation	
Trench excavation—hand (to 4' deep)	2.40
Pipe trench—backfill and compaction, loader and compactors, and hand and compactors	
Backfill and compaction	MH/CY
Backfill and compaction—loader and compactors	0.60
Backfill and compaction—hand and compactors	3.40
Pipe trench—sand bedding and shading with loader	
	MH/CY
Sand bedding and shading with loader	0.40
Pipe trench—shoring and bracing trenches	
Shoring (place/remove)	MH/SF
Laborer	5.50
Carpenter	3.00

2.9 Schedule G—Simple foundations for industrial plants

Standard labor estimating units

Facility—industrial plant (simple foundation)

	MH/CY	MH/SF	MH/LF	MH/LB	MH/DIA IN FT
Structure excavation—backhoe	0.20				
Structure excavation—hand	2.80				
Structure backfill and compact— loader and wacker	0.60				
Structure backfill and compact— hand	3.00				
Edge forms—slabs and foundations			0.10		
Fabricate, install, strip foundation forms—one use		0.30			
Fabricate, install, strip pedestal forms—one use		0.25			
Fabricate, install, strip wall forms—one use		0.20			
Fabricate and install reinforcing steel				0.03	
Layout templets and set anchor bolts					0.25
Set embedded steel—crub angle, etc.				0.04	
Place concrete—from truck below grade	0.80				
Place concrete—slabs at grade	1.00				
Place concrete—pedestals and walls	2.00				
Finish flate concrete		0.12			
Patch and sack concrete		0.05			
Install mesh			0.02		

2.10   Schedule M—Pipe supports and hangers

Facility—Industrial plant, pipe supports, and hangers

Pipe support spacing	Span in feet	
Pipe size (inches)	Water service	Steam gas and air
1	7	9
1.5	9	12
2	10	13
2.5	11	14
3	12	15
3.5	13	16
4	14	17
5	16	19
6	17	21
8	19	24
10	22	28
12	23	30
14	25	32
16	27	35
18	28	37
20	30	39
24	32	42
Standard labor estimating units		
Type of hanger	MH	
Pipe size (1″–4″)		
Clevis, band, ring, and expansion hangers	3.8	
Trapeze hanger	6.0	
Pipe size		
6″	5.0	
8″	5.3	
10″	5.6	
12″	6.0	
Shoes and guides	2.0	
Engineered supports, spring, sway, snubbers	1.0 MH/DI	
Structural supports (angles, channels, beams, tube steel)—0.04 per LB		

2.11 Standard and line pipe—Wall thickness

Pipe size	40	STD	60	80	XH	100	120	140	160	XXH
0.5	0.109	0.109		0.147	0.147				0.188	0.294
0.75	0.113	0.113		0.154	0.154				0.219	0.308
1	0.133	0.133		0.179	0.179				0.250	0.358
1.5	0.145	0.145		0.200	0.200				0.281	0.400
2	0.154	0.154		0.218	0.218				0.344	0.436
2.5	0.203	0.203		0.276	0.276				0.375	0.552
3	0.216	0.216		0.300	0.300				0.438	0.600
4	0.237	0.237		0.337	0.337		0.438		0.531	0.674
6	0.280	0.280		0.432	0.432		0.562		0.719	0.864
8	0.322	0.322	0.406	0.500	0.500	0.594	0.719	0.812	0.906	0.875
10	0.365	0.365	0.500	0.594	0.500	0.719	0.844	1.000	1.125	1.000
12	0.406	0.375	0.562	0.688	0.500	0.844	1.000	1.125	1.312	1.000
14	0.438	0.375	0.594	0.750	0.500	0.938	1.094	1.250	1.406	
16	0.500	0.375	0.656	0.844	0.500	1.031	1.219	1.438	1.594	
18	0.562	0.375	0.750	0.938	0.500	1.156	1.375	1.562	1.781	
20	0.594	0.375	0.812	1.031	0.500	1.281	1.500	1.750	1.969	
22		0.375	0.875	1.125	0.500	1.375	1.625	1.875	2.125	
24	0.688	0.375	0.969	1.219	0.500	1.531	1.812	2.062	2.344	

Chapter 3

Oil refinery

3.1 Section introduction

This section provides the reader a basic understanding of the fundamentals and the operating relationship between the plant equipment in an oil refinery and covers the craft labor for the assembly and field erection required to put the equipment into operation in an oil refinery. The oil refinery or petroleum refinery is an industrial process plant where crude oil is transformed and refined into more useful products. Oil refineries are large industrial complexes with extensive piping running throughout, carrying product between chemical processing units. The crude oil feedstock has been processed by an oil production plant. There is an oil depot at or near an oil refinery for the storage of incoming crude oil feedstock and bulk liquid products. Petroleum refineries have many different processing units and auxiliary facilities. Each refinery has its own arrangement and combination of refining process determined by the location, desired products, and economic considerations.

3.2 Stripper unit

3.2.1 Scope of field work required for stripper unit

Scope of work-field erection

Column—T-0000 stripper

Drums

Feed coalescer
Feed surge drum
Cold high-pressure separator
Cold low-pressure separator
Stripper overhead receiver
Fuel gas knockout drum

Industrial Process Plant Construction Estimating and Man-Hour Analysis.
https://doi.org/10.1016/B978-0-12-818648-0.00003-X

Heaters

Combined feed heater
Stripper reboiler

Heat exchangers

Cold combined feed exchanger
Hot combined feed exchanger
Stripper/reactor effluent exchanger
Diesel product/stripper feed exchanger

Air coolers

Diesel product cooler
Reactor effluent condenser
Stripper overhead condenser

Pumps

Feed charge pump (centrifugal)
Wash water pump (centrifugal)
Stripper overhead pump (centrifugal)
Stripper bottoms pump (centrifugal)
Stripper overhead water pump (centrifugal)

Compressors

Recycle gas compressor
Lube oil console
Lube oil tank
Seal oil console
Main LO pump
Aux LO pump
Equipment monitoring system

Filters—Fuel gas filter
Reactors

Diolefin reactor
Hydrotreating reactor

3.3 Stripper unit estimating data

3.3.1 Vessels/columns—Pressure equipment sheet 1

Description	Weight range tons	MH	Unit
Vertical vessels (towers)			
Stripper ID 7′ × T/T 116′ 55 ton	41–60	9.45	Ton
84″ Install double downflow valve trays		30	EA
84″ Install demisting pads (single grid support, pad, and grid top)		48	EA
Vortex breaker		32	EA
Remove and replace manway cover (24″ 300# removable-Davit)		40	EA
Install platforms and ladders		40	Ton

3.3.2 Horizontal vessels—Pressure equipment sheet 2

Description	Weight range tons	MH	Unit
Drums			
Feed coalesce ID 8′ × T/T 16′ 7.75 ton	6–10	12.0	Ton
Feed Surge Drum IF 8.5′ × 23′ 15.25 ton	11–20	10.0	Ton
Cold high-pressure separator ID 10′ × 30′ 115 ton	101–150	4.1	Ton
Cold low-pressure separator ID 8.5′ × 28′ 33.6 ton	31–40	8.1	Ton
Stripper overhead receiver ID 7.5′ × 24′ 12 ton	11–20	10.0	Ton
Fuel gas knockout drum ID 1.5′ × 6.5	0–5	18.0	Ton

3.3.3 Heaters, heat exchangers, and air coolers sheet 3

Description	Weight range tons	MH	Unit
Combined feed heater footprint-ft × ft 12 × 12 45 ton	41–60	7.0	Ton
Stripper reboiler footprint-ft × ft 14.5 × 14.5 7.5 ton	6–10	12.0	Ton
Cold combined feed exchanger ID-ft 2.5 T/T-ft 17.5 8 ton	6–10	12.0	Ton
Hot combined feed exchanger ID-ft 3.6 T/T-ft 20 14.5 ton	11–20	10.0	Ton
Stripper/reactor effluent exchanger ID-ft 2.9 T/T-ft 17.5 12.5 ton	11–20	10.0	Ton
Diesel product/stripper feed exchanger ID-ft 1.6 T/T-ft 20 4.5 ton	0–5	18.0	Ton
Diesel product cooler T/T-ft 30 ft × ft 15.5 × 30 25.5 ton	21–30	8.3	Ton
Reactor effluent condenser T/T-ft 30 ft × ft 56 × 34 124 ton	101–150	4.1	Ton
Stripper overhead condenser T/T 30 ft × ft 26 × 30 32.5 ton	31–40	8.1	Ton

3.3.4   Pump—Centrifugal sheet 4

Description	HP range	MH	Unit
Feed charge pump (centrifugal) 6′ × 4′ × 4′ 7.5 ton HP 600	501–5000	0.60	HP
Wash water pump (centrifugal) 6 × 4.5 × 3.5 1.5 ton HP 40	31–50	2.00	HP
Stripper overhead pump (centrifugal) 8 × 3 × 1.8 ton HP 100	76–100	1.50	HP
Stripper bottoms pump (centrifugal) 9 × 3.5 × 3.5 2 ton HP 200	126–300	1.25	HP
Stripper overhead water pump (centrifugal) 6 × 3 × 3 0.75 ton HP 0.75	0–1	24.00	HP

3.3.5   Filter and reactor sheet 5

Description	Weight range tons	MH	Unit
Fuel gas filter 0.2 ton	1	24	EA
Diolefin reactor ID-ft 9 T/T-ft 16 65 ton	61–100	5.5	Ton
Hydrotreating reactor ID-ft 9 T/T-ft 80 276 ton	201–250	3.2	Ton

3.4   Stripper unit estimate

3.4.1   Vessels/columns—Pressure equipment sheet 1

Description	Historical			Estimate		
	MH	Qty	Unit	Qty	Unit	BM
Stripper vertical tower						0
Stripper ID 7′ × T/T 116′ 55 ton	9.45	55.0	Ton	0	Ton	0
Install double downflow valve trays	30.0	16.0	EA	0	EA	0
Install demisting pads (single grid support, pad, and grid top)	48.0	1.0	EA	0	EA	0
Vortex breaker	32.0	1.0	EA	0	EA	0
Remove and replace manway cover (24″ 300# Removable-Davit)	40.0	1.0	EA	0	EA	0
Install platforms and ladders	40.0	1.5	Ton	0	Ton	0

3.4.2   Horizontal vessels—Pressure equipment sheet 2

Description	Historical			Estimate		
	MH	Qty	Unit	Qty	Unit	BM
Drums						**0**
Feed coalesce ID 8′ × T/T 16′ 7.75 ton	12.0	7.8	Ton	0	Ton	0
Feed surge drum IF 8.5′ × 23′ 15.25 ton	10.0	15.3	Ton	0	Ton	0
Cold high-pressure separator ID 10′ × 30′ 115 ton	4.1	115.0	Ton	0	Ton	0
Cold low-pressure separator ID 8.5′ × 28′ 33.6 ton	8.1	33.6	Ton	0	Ton	0
Stripper overhead receiver ID 7.5′ × 24′ 12 ton	10.0	12.0	Ton	0	Ton	0
Fuel gas knockout drum ID 1.5′ × 6.5′	18.0	1.0	Ton	0	Ton	0

3.4.3   Heaters, heat exchangers, and air coolers sheet 3

Description	Historical			Estimate		
	MH	Qty	Unit	Qty	Unit	BM
Heaters, heat exchangers, and air coolers						**0**
Combined feed heater footprint-ft × ft 12 × 12	7.0	45.0	Ton	0	Ton	0
Stripper Reboiler Footprint-ft 14.5 × 14.5	12.0	7.5	Ton	0	Ton	0
Cold combined feed exchanger ID-ft 2.5 T/T-ft 17.5	12.0	8.0	Ton	0	Ton	0
Hot combined feed exchanger ID-ft 3.6 T/T-ft 20	10.0	14.5	Ton	0	Ton	0
Stripper/reactor effluent exchanger ID-ft 2.9 T/T-ft 17.5	10.0	12.5	Ton	0	Ton	0
Diesel product/stripper feed exchanger ID-ft 1.6 T/T-ft 20	18.0	4.5	Ton	0	Ton	0
Diesel product cooler T/T-ft 30 ft × ft 15.5 × 30	8.3	25.5	Ton	0	Ton	0
Reactor effluent condenser T/T-ft 30 ft × ft 56 × 34	4.1	124.0	Ton	0	Ton	0
Stripper overhead condenser T/T 30 ft × ft 26 × 30	8.1	32.5	Ton	0	Ton	0

3.4.4   Pump—Centrifugal sheet 4

Description	Historical			Estimate		
	MH	Qty	Unit	Qty	Unit	BM
Pump—centrifugal						**0**
Feed charge pump (centrifugal)	0.60	600.0	HP	0	HP	0
$6' \times 4' \times 4'$ 7.5 ton						
Wash water pump (centrifugal)	2.00	40.0	HP	0	HP	0
$6 \times 4.5 \times 3.5$ 1.5 ton						
Stripper overhead pump (centrifugal)	1.50	100.0	HP	0	HP	0
$8 \times 3 \times 3$ 1.8 ton						
Stripper bottoms pump (centrifugal)	1.25	200.0	HP	0	HP	0
$9 \times 3.5 \times 3.5$ 2 ton						
Stripper overhead water pump (centrifugal)	24.00	0.8	HP	0	HP	0
$6 \times 3 \times 3$ 0.75						

3.4.5   Filter and reactor sheet 5

Description	Historical			Estimate		
	MH	Qty	Unit	Qty	Unit	BM
Filter and reactor						**0**
Fuel gas filter	24	1	EA	0	EA	0
Diolefin reactor ID-ft 9 T/T-ft 16	5.5	65	Ton	0	Ton	0
Hydrotreating reactor ID-ft 9 T/T-ft 80	3.2	276	Ton	0	Ton	0

3.4.6 Recycle compressor reciprocating sheet 6

Description	MH	Historical Qty	Unit	Estimate Qty	Unit	BM
Recycle compressor reciprocating						**0**
Frame and Gear	340	1	EA	0	EA	0
34'-5" Cylinder	180	1	EA	0	EA	0
22'-0" Cylinder	120	1	EA	0	EA	0
First stage suction damper	60	1	EA	0	EA	0
First stage discharge damper	60	1	EA	0	EA	0
Extended bearing and postal	60	1	EA	0	EA	0
Lube oil consul 2 ea.	60	2	EA	0	EA	0
Moister separator 3 ea.	60	3	EA	0	EA	0
Full main tech deck	240	1	Lot	0	Lot	0
IC Pipe—Interstate piping spools between coolers	800	1	Lot	0	Lot	0
Rotor and extension shaft	240	1	EA	0	EA	0
Motor stator	180	1	EA	0	EA	0
Motor enclosure (soleplates and Nc)	200	1	Lot	0	Lot	0

3.5 Stripper unit—Equipment installation man-hours

Facility-Oil Refinery	Actual MH	Estimated BM
Stripper vertical tower	1180	0
Drums	1127	0
Heaters, heat exchangers, and air coolers	1835	0
Pump—centrifugal	858	0
Filter and reactor	1265	0
Recycle compressor reciprocating	2780	0
Equipment installation man-hours	9045	0

Chapter 4

Waste heat boiler

4.1 Equipment descriptions

4.1.1 Boiler pressure parts

The section includes the installation of a new waste heat boiler.

Boiler-pressure parts
The furnace tubes and generating bank tubes will be shipped knocked down with as much factory assembly as practical for shipping.

Steam drum and mud drum
Install one (1) steam drum and one (1) mud drum. The steam drum will come with all internals mounted. Remove and reinstall after boiling out.

Furnace panels
The furnace consists of wall panels and roof panels.

Water wall panels
Install water wall panels.

Coen 100 MMBTU burner with primary air fan
Set and seal weld the burner wind box to the casing on the furnace. Set IC duct-work from the wind box to the new fan including expansion joint and silencer. New fan will be mounted on the structural steel. Set all loose spool piping assemblies including the jackshaft assemblies. Fabricate and install silencer support.

Install super heater section
Connection to drums
Generating bank, boiler, sidewall, furnace, roof, and furnace rear wall tubes are expanded and flared to 1800 psi.

Soot blowers
Install rotary and retractable soot blowers and support bearings.

Fuel feeding equipment
Metering bins
Install six (6) metering bin screws. Installation includes support steel. Metering bins will set on load cells. Install 24 load cells.

Boiler drag chain
Install one (1) boiler drag chain above the metering bins.

Industrial Process Plant Construction Estimating and Man-Hour Analysis.
https://doi.org/10.1016/B978-0-12-818648-0.00004-1

Install $8'\text{-}0'' \times 6'\text{-}0''$ enclosure and six (6) rack and pinions, slide gates, including chain wheel actuators between the meters.

Solid fuel is fed to the boiler in six (6) feed points, all on the front wall.

Fuel chutes

Install metering bin discharge chutes, expansion joints to stoker windswept spouts, support bins, and drag chains for all six (6) bins.

4.1.2   Structural steel and boiler casing

Structural steel

Erect structural steel for:
   Stoker support steel
   Boiler
   Economizer
   Mechanical collector
   Air heater
   Scrubber
   Stack

Boiler casing

Erect casing—generating bank, boiler, and penthouse

4.2 General scope of field work required for each 400,000 lb./h waste heat boiler 400,000 lb./h membrane wall boiler with superheater scope of work-field erection

Pressure parts

Install steam and mud drums
Steam drum trim piping
Drum internals—remove and reinstall
Generating bank tubes
Install generating tubes
2-1/2″ OD × 0.203″ Wall thickness, swage, and roll
Ground assembly—headers, furnace, and water wall panels
Assemble headers and wall panels, fit and weld water wall panels to headers
Erect headers and wall panels, fit and weld water wall panels
Erect side wall, front, and rear panels—water wall; 2-31/32″ OD × 0.203″ WT, BW (TIG)
Fit and weld filler bar at tube welds
Fit and weld filler joining membrane panels
Fit and weld water wall tubes; 2-31/32″ OD × 0.203″ WT, BW (TIG)
Erect, fit, and weld primary/secondary headers and superheater elements
Primary superheater headers/coils
10.75″ Diameter inlet header
20″ Outlet header
Weld superheater tubes; 1.772″ OD × 0.150″ WT, BW (TIG)
Burner system
Burner and wind box
Fan with drive
Silencer
Soot blowers
Install rotary and retractable soot blowers and support bearings
Erect and install down comers and steam code piping
Scope of Field Work Required For Structural Steel and Boiler Casing
Erect boiler structural steel
    Column line "1" (west)
    Column line "6" (east)
    Column line "7" (east)
    Column line "A" (north)
    Column line "B" (north)
    Column line "F" (south)
    Column line "G" (south)
Install platform, grating, and handrail @ El. 100′-0″ to 184′-0″
Install stair nos. 1 and 2
Erect penthouse steel
Erect casing—generating bank, boiler, and penthouse

4.3   400,000 lb./h waste heat boiler estimate data

4.3.1   Drums, generating tubes, headers, side, front and rear wall panels sheet 1

Description	MH	Unit
Drums—includes straps and/or U-bolts		
100,001–200,000 lbs	4.60	Ton
200,001–300,000 lbs	3.80	Ton
Steam drum trim piping	260.00	LOT
Drum internals—remove and reinstall	136.00	Boiler
Generating bank tubes		
Install generating tubes	0.50	EA
2'1/2" Ends—expand tubes in steam and mud drums	0.44	End
Headers, furnace, and water wall panels		
Headers—loose	4.00	Ton
Shop assembled wall panels with or without headers attached	2.50	Ton
Field tube welding—over 2-1/2" and including 3" TIG (1500 PSI)	4.20	EA
Primary super heater headers/coils	4.20	Ton
Fit and weld filler bar at tube welds	1.00	Space
Fit and weld filler joking membrane panels	0.60	LF
Field tube welding—over 1-1/2" and including 2" TIG (1500 PSI)	3.70	EA

4.3.2   Erect headers and panels, weld water wall tubes, burner, soot blower, superheater sheet 2

Description	MH	Unit
Erect headers and wall panels, fit and weld water wall panels		
Erect right, left front, rear, and roof panels	2.50	Ton
Fit and weld filler bar at tube welds	1.00	Space
Fit and weld filler joining membrane panels	0.60	LF
Scaffolding and rigging	136.00	Boiler
Fit and weld water wall tubes; 2-31/32" OD × 0.203" WT, BW (TIG)	4.20	EA
Erect, fit, and weld primary/secondary headers and superheated elements		
Primary superheater headers/coils (27.5' L × 10' W × 10.5 H)	4.20	Ton
10.75" Diameter inlet header	4.00	Ton
20" Outlet header	4.00	Ton
Weld super heater tubes; 1.772" OD × 0.150" WT, BW (TIG)	3.70	EA
Burner system		
Burner and wind box	59.40	Ton
Fan with drive	45.00	Ton
Silencer	60.00	EA
Soot blower (24' L × 1'-6" diameter × 2'-0" H 2,570 lb. per unit	64.00	EA

4.3.3 Down comers and code piping sheet 3

Description	MH	Unit
Erect and install down comers and code piping		
Handle down comer; 12″ OD × 0.562″ WT	1.32	LF
12″ × 0.562 WT CS BW	12.6	EA
Handle 12″ 300# BW valve	5.4	EA
Handle 10″ 300# BW valve	4.50	EA
Handle 8″ 300# BW spray control valves	7.20	EA
Handle 3″ 300# safety/control valves	2.70	EA
Handle 2″ 300# control valves—boiler	1.00	EA
10″ × 0.562 WT CS BW	10.5	EA
8″ × 0.562 WT CS BW	8.40	EA
3″ × 0.375 WT CS BW	1.50	EA
2″ × 0.375 WT CS BW	1.10	EA
4″ 300# Flanged air flow elements	3.60	EA
Handle 10″ × 0.562 pipe	1.10	LF
Handle 8″ × 0.562 pipe	0.88	LF
4″ 300# Bolt connection	1.60	EA

4.4   Waste heat boiler installation estimate

4.4.1   Drums, generating tubes, headers, side, front and rear panels sheet 1

Description	Estimate					
	MH	Qty	Unit	Qty	Unit	BM
Install steam and mud drums						**0**
Steam drum (6'-0" diameter × 48')	3.8	111	Ton	0	Ton	0
Mud drum (4'-6" diameter × 47')	4.6	63	Ton	0	Ton	0
Steam drum trim piping	260.0	1	LOT	0	LOT	0
Drum internals—remove and reinstall	136.0	1	Boiler	0	Boiler	0
Generating bank tubes						**0**
2-1/2" OD × 0.203" Wall thickness, swage, and roll						
Install generating tubes	0.50	2629	EA	0	EA	0
2'1/2" Ends—expand tubes in steam and mud drums	0.44	5258	End	0	End	0
Ground assembly—headers, furnace, and wall panels						**0**
Fit and weld water wall panels to headers						
Lower side wall header—lower side wall panels						
Place lower side wall header	4.0	11	Ton	0	Ton	0
Place lower side wall panels—water wall	2.5	47	Ton	0	Ton	0
Field tube welding—over 2-1/2" TIG	4.2	156	EA	0	EA	0
Upper side wall header—upper side wall panels						
Place upper side wall header	4.0	9	Ton	0	Ton	0
Place upper side wall panels—water wall	2.5	47	Ton	0	Ton	0
Field tube welding—over 2-1/2" TIG	4.2	156	EA	0	EA	0
Place front wall header	4.0	5	Ton	0	Ton	0
Place front wall panels	2.5	24	Ton	0	Ton	0
Field tube welding—over 2-1/2" TIG	4.2	108	EA	0	EA	0
Place rear wall header	4.0	4	Ton	0	Ton	0
Place rear wall panels—water wall	2.5	24	Ton	0	Ton	0
Field tube welding—over 2-1/2" TIG	4.2	108	EA	0	EA	0

4.4.2 Erect headers and panels, weld water wall tubes, burner, soot blower, superheater sheet 2

Description	Historical			Estimate		
	MH	Qty	Unit	Qty	Unit	BM
Erect headers and wall panels, fit and weld wall panels						**0**
Scaffolding and rigging	136.00	1.0	Boiler	0	Boiler	0
Erect right side wall panels—water wall	2.50	120.0	Ton	0	Ton	0
Fit and weld filler bar at tube welds	1.00	468.0	Space	0	Space	0
Fit and weld filler joining membrane panels	0.60	160.0	LF	0	LF	0
Erect left side wall panels—water wall	2.50	120.0	Ton	0	Ton	0
Fit and weld filler bar at tube welds	1.00	468.0	Space	0	Space	0
Fit and weld filler joining membrane panels	0.60	160.0	LF	0	LF	0
Erect front wall panels—water wall	2.50	120.0	Ton	0	Ton	0
Fit and weld filler bar at tube welds	1.00	432.0	Space	0	Space	0
Fit and weld filler joining membrane panels	0.60	156.0	LF	0	LF	0
Erect rear wall panels—water wall	2.50	120.0	Ton	0	Ton	0
Fit and weld filler bar at tube welds	1.00	432.0	Space	0	Space	0
Fit and weld filler joining membrane panels	0.60	156.0	LF	0	LF	0
Erect roof panels—water wall	2.50	90.0	Ton	0	Ton	0
Fit and weld filler bar at tube welds	1.00	588.0	Space	0	Space	0
Fit and weld filler joining membrane panels	0.60	173.0	LF	0	LF	0
Fit and weld water wall tubes	4.20	1645.8	EA	0	EA	**0**
Erect, fit, and weld primary/ secondary headers						**0**
Primary superheater headers/coils	4.20	96.6	Ton	0	Ton	0
10.75″ Diameter inlet header	4.00	2.1	Ton	0	Ton	0
20″ Outlet header	4.00	5.0	Ton	0	Ton	0
Weld superheater tubes; 1.772″ OD	3.70	292.0	EA	0	EA	0
Burner System						**0**
Burner and wind box	59.40	4.7	Ton	0	Ton	0
Fan with drive	0.00	1.8	Ton	0	Ton	0
Silencer	0.00	1.0	EA	0	EA	0
Soot blowers						**0**
Soot blower						
Install rotary and retractable soot blowers						
Superheater—soot blower with pipe, valve, fitting	64.00	12.0	EA	0	EA	0
Economizer—soot blower with pipe, valve, fitting	64.00	4.0	EA	0	EA	0
Generating bank—soot blower with PVF	64.00	4.0	EA	0	EA	0

4.4.3   Down comers and code piping sheet 3

Description	Historical			Estimate		
	MH	Qty	Unit	Qty	Unit	BM
Erect and install down comers and code piping						0
Handle down comers; 12″ OD × 0.562 WT × 40′	1.32	280.0	LF	0	LF	0
Down comers; 12″ OD × 0.562 WT BW	12.6	14.0	EA	0	EA	0
Handle 12″ 300# BW main stop valve and Ck valve	5.4	2.0	EA	0	EA	0
12″ × 0.562 WT CS BW	12.60	4.0	EA	0	EA	0
Handle 10″ 300# BW feed water stop and Ck valve	136.00	2.0	EA	0	EA	0
10″ × 0.562 WT CS BW	4.00	4.0	EA	0	EA	0
Handle 8″ 300# BW spray control valves	7.20	4.0	EA	0	EA	0
8″ × 0.562 WT CS BW	8.40	8.0	EA	0	EA	0
Handle 3″ 300# safety valves	2.70	3.0	EA	0	EA	0
3″ × 0.375 WT CS BW	1.50	6.0	EA	0	EA	0
Handle 3″ 300# control valves—boiler	2.70	2.0	EA	0	EA	0
3″ × 0.375 WT CS BW	1.50	4.0	EA	0	EA	0
Handle 2″ 300# control valves—boiler	1.00	2.0	EA	0	EA	0
2″ × 0.375 WT CS BW	1.10	4.0	EA	0	EA	0
Handle 4″ 300# flanged flow elements	3.60	6.0	EA	0	EA	0
4″ 300# Bolt up	1.60	12.0	EA	0	EA	0
Handle 10″ × 0.562 pipe	1.10	160.0	LF	0	LF	0
Handle 8″ × 0.562 pipe	0.88	160.0	LF	0	LF	0
Handle 4″ × 0.375 pipe	0.28	225.0	LF	0	LF	0
Handle 3″ × 0.375 pipe	0.21	160.0	LF	0	LF	0
Handle 2″ × 0.375 pipe	0.18	160.0	LF	0	LF	0

4.5   Waste heat boiler—Equipment installation man-hours

Facility—biomass plant major equipment	Actual	Estimated
	MH	BM
Install steam and mud drums	1103	0
Generating bank tubes	3628	0
Ground assembly—headers, furnace, and water wall panels	2692	0
Erect headers and wall panels, fit and weld water wall panels	4432	0
Fit and weld water wall tubes; 2-31/32″ OD × 0.203″ WT, BW (TIG)	6912	0
Erect, fit and weld primary/secondary headers and superheater elements	1514	0
Burner system	280	0
Soot blowers	1280	0
Erect and install down comers and code piping	1509	0
Waste heat boiler—equipment and piping installation man-hours	23351	0

4.6 Scope of field work required for structural steel and boiler casing

Erect boiler structural steel
    Column line "1" (west)
    Column line "6" (east)
    Column line "7" (east)
    Column line "A" (north)
    Column line "B" (north)
    Column line "F" (south)
    Column line "G" (south)
Install platform, grating, and handrail @ El. 100'-0" to 184'-0"
Install stair nos. 1 and 2
Erect penthouse steel
Erect casing—generating bank, boiler, and penthouse

4.7 Structural steel and boiler casing estimate data

4.7.1 Structural steel sheet 1

Description	MH	Unit
Erect boiler structural steel (stairs, platforms, grating, and handrails)		
Main steel	24.00	Ton
Platform framing	0.15	SF
Grating	0.20	SF
Handrail	0.25	LF
Erect penthouse steel	24.00	Ton
Install platform, grating, and handrail @ El. 100'-0" to 184'0"		
Platform framing	0.15	SF
Grating	0.20	SF
Handrail	0.25	LF
Install stair nos. 1 and 2		
Structural steel	24.00	Ton
Platform framing	0.15	SF
Grating	0.20	SF
Handrail	0.25	LF
Stair treads	0.85	EA

4.7.2   Boiler casing sheet 2

Description	MH	Unit
Erect casing—generating bank, boiler, and penthouse	62	Ton

4.8   Structural steel installation estimate

4.8.1   Structural steel sheet 1

	Historical			Estimate		
	MH	Qty	Unit	Qty	Unit	BM
Erect boiler structural steel						0
Column line "1" (west)	24.00	51.1	Ton	0	Ton	0
Column line "6" (east)	24.00	10.7	Ton	0	Ton	0
Column line "7" (east)	24.00	10.7	Ton	0	Ton	0
Column line "A" (north)	24.00	14.0	Ton	0	Ton	0
Column line "B" (north)	24.00	37.9	Ton	0	Ton	0
Column line "F" (south)	24.00	30.2	Ton	0	Ton	0
Column line "G" (south)	24.00	37.2	Ton	0	Ton	0
Erect penthouse steel	24.00	45.8	Ton	0	Ton	0
Install platform, grating, and handrail @ El. 100′-0″ to 184′-0″						0
Platform framing	0.15	15488.0	SF	0	SF	0
Grating	0.20	15488.0	SF	0	SF	0
Handrail	0.25	2816.0	LF	0	LF	0
Install stair nos. 1 and 2						0
Structural steel	24.00	14.4	Ton	0	Ton	0
Platform framing	0.15	1440.0	SF	0	SF	0
Grating	0.20	1440.0	SF	0	SF	0
Handrail	0.25	320.0	LF	0	LF	0
Stair treads	0.85	192.0	EA	0	EA	0

4.8.2   Boiler casing sheet 2

	Installation man-hour					
		Historical			Estimate	
	MH	Qty	Unit	Qty	Unit	BM
Erect casing—generating bank, boiler, and penthouse	62	19.0	Ton	0	Ton	**0**

4.9 Structural steel and boiler casing-equipment installation man hours

Facility—biomass plant major equipment	Actual MH	Estimated BM
Erect boiler structural steel	4601	0
Erect penthouse steel	1100	0
Install platform, grating, and handrail @ El. 100′-0″ to 184′-0″	6140	0
Install stair Nos. 1 and 2	1094	0
Erect casing—generating bank, boiler, and penthouse	1180	0
Structural steel and boiler casing-equipment installation man-hours	**14116**	**0**

Chapter 5

Pulp and paper

5.1 Recovery boiler

Design features:
A chemical and heat recovery boiler for the pulp and paper industry: Concentrated black liquor and combustion air are introduced into the furnace where heat is recovered as steam for electricity generation and process heating, and the inorganic portion of the black liquor is recovered as sodium compounds.

Liquor processing capacity:
To 10,000,000 lb/day; (4500 t/day) dry solids

Steam pressure:
To 1850 psi (12.6 map) design

Steam temperature:
To 950°F (510°C)

Fuel:
Pulp mill liquor

5.2 Reheat recovery boiler

Design features:
A chemical and heat recovery boiler for the pulp and paper industry with reheat and optional dual pressure configuration to improve lower furnace corrosion protection.

Steam pressure:
To 2600 psi (17.9 map)

Steam temperature:
To 950°F (510°C)

Fuel:
Pulp mill liquor

Industrial Process Plant Construction Estimating and Man-Hour Analysis.
https://doi.org/10.1016/B978-0-12-818648-0.00005-3

5.3 General scope of field work required for each recovery boiler sheet 1

5.3.1 Scope of work-field erection

Pressure parts

Install steam drum
Drums include straps and/or U-bolts: steam drum
Steam drum trim piping
Drum internals—remove and reinstall
Generating bank
Generating tube sections and headers—unload and handle
Set rigging for hoisting assemblies
Hoist into place
Weld tubes to header stubs—2-1/2″ diameter, TIG weld
Fit and weld filler bar at tube welds
Integral economizers for recovery units—multiple header type
Economizer tube sections and headers—unload and handle
Set rigging for hoisting assemblies
Hoist into place
Weld tubes to header stubs—2-1/2″ diameter, TIG weld
Fit and weld filler bar at tube welds
Baffles—place and weld including clips
Vibration bars
Backstays
Casing and hoppers
Doors
Headers, furnace, and waterwall panels
Boiler sidewall tubes
Wall panels—shop assembled with headers
Fit and weld filler joining membrane panels
Furnace front wall tubes
Headers—loose
Wall panels—shop assembled without headers
Weld tubes to header stubs—2-1/2″ diameter, TIG weld
Fit and weld filler bar at tube welds
Fit and weld filler joining membrane panels
Furnace roof tubes
Headers—loose
Tubes—loose, furnace including screen
Weld tubes to header stubs—3″ diameter, TIG weld
Furnace sidewall tubes
Headers—loose
Wall panels—shop assembled without headers
Weld tubes to header stubs—2-1/2″ diameter, TIG weld
Fit and weld filler bar at tube welds
Fit and weld filler joining membrane panels

5.4   General scope of field work required for each recovery boiler sheet 2 scope of work-field erection

Pressure Parts

Furnace rear wall tubes
Furnace rear wall tubes (wall panel—membrane)
Furnace rear wall tubes (loose)
Weld tubes to header stubs—2-1/2″ diameter, TIG weld
Fit and weld filler bar at tube welds
Fit and weld filler joining membrane panels
Superheaters for recovery boilers
Tubes—connecting
Tubes—sections
Headers
Weld superheater tubes; 1.77″ OD × 0.150″ WT, BW (TIG)
Attemperators for recovery units
Spray type—complete with body and internals (excludes welding)
Drum type—complete with internals, piping, valves, and supports (excludes welding)
Supply and riser tubes
Headers
Tubes—makeup or supply
Tubes—riser
Weld tubes to header stubs—4-1/2″ diameter, TIG weld
Downcomers
Install downcomer
Diameter weld, girth
Diameter weld, ring
Diameter weld, ring
Smelt spout and hood
Auxiliary equipment
Floor and hopper support steel
Suspension steel
Buckstays
Casing
Burner wind box—primary, secondary, and tertiary
Primary burners
Lighters
Hoppers
Doors and frames
Seals
Expanded metal lathe
Closure plates
Instrument connection

5.5 Recovery boiler estimate data

5.5.1 Erect drum, generating bank, integral economizers multiple header type sheet 1

Description	MH	Unit
Drums include straps and/or U-bolts		
100,001–200,000 lbs	4.60	Ton
200,001–300,000 lbs	3.80	Ton
Steam drum trim piping	260.00	LOT
Drum internals—remove and reinstall	136.00	Boiler
Generating bank		
Generating tube sections and headers—unload and handle	10.00	Ton
Set rigging for hoisting assemblies	60.00	Boiler
Hoist into place	1.50	Ton
Weld tubes to header stubs—2-1/2″ diameter, TIG no PWHT	2.40	EA
Fit and weld filler bar at tube welds	1.00	Space
Integral economizers multiple header type		
Economizer tube sections and headers—unload and handle	10.00	Ton
Set rigging for hoisting assemblies	60.00	Boiler
Hoist into place	1.50	Ton
Weld tubes to header stubs—2-1/2″ diameter, TIG no PWHT	2.40	EA
Fit and weld filler bar at tube welds	1.00	Space
Baffles—place and weld including clips	60.00	Ton
Vibration bars	65.00	Ton
Buckstays	40.00	Ton
Casing and hoppers	55.00	Ton
Doors	55.00	Ton

5.5.2 Erect headers and panels, furnace wall and roof tubes, weld tubes and panels sheet 2

Description	MH	Unit
Erect headers and wall panels, fit and weld waterwall panels		
Boiler sidewall, furnace front wall, roof, sidewall and rear wall tubes		
Wall panels—shop assembled with headers	10.00	Ton
Fit and weld filler joining membrane panels	0.70	LF
Headers—loose	15.00	Ton
Wall panels—shop assembled without headers	10.00	Ton
Weld tubes to header stubs—2-1/2″ diameter, TIG no PWHT	2.40	EA
Headers—loose	30.00	Ton
Tubes—loose, furnace including screen	30.00	Ton
Weld tubes to header stubs—3″ diameter, TIG no PWHT	3.00	EA
Furnace rear wall tubes (wall panel—membrane)	12.00	Ton
Furnace rear wall tubes (loose)	30.00	Ton
Fit and weld filler bar at tube welds	1.00	Space

5.5.3   Erect/weld super heater, supply and riser tube, down comers, attemperators, sheet 3

Description	MH	Unit
Superheater, supply and riser tubes, and downcomers		
Tubes—connecting	32.00	Ton
Tubes—sections	10.00	Ton
Headers	15.00	Ton
Weld superheater tubes; 1.77″ OD × 0.150″ WT, BW (TIG)	2.10	EA
Attemperators for recovery units		
Spray type—complete with body and internals (excludes welding)	20.00	Ton
Drum type—complete with internals, piping, valves, and supports (excludes welding)	45.00	Ton
Headers	15.00	Ton
Tubes—make up or supply	25.00	Ton
Tubes—riser	29.00	Ton
Weld tubes to header stubs—4-1/2″ diameter, TIG no PWHT	4.00	EA
Install downcomer	10.00	Ton
Downcomers; 10″ OD × 0.500 WT BW	0.55	DI
Downcomers; 12″ OD × 0.562 WT BW	1.05	DI
Downcomers; 14″ OD × 0.594 WT BW	1.05	DI

5.5.4   Install smelt spout and hood and auxiliary boiler support steel and equipment, sheet 4

Description	MH	Unit
Smelt spout and hood	55.00	Ton
Auxiliary boiler support steel and equipment		
Floor and hopper support steel	28.00	Ton
Suspension steel	30.00	Ton
Buckstays	36.00	Ton
Casing	95.00	Ton
Burner wind box—primary, secondary, and tertiary	75.00	Ton
Primary burners	72.00	EA
Lighters	5.00	EA
Hoppers	95.00	Ton
Doors and frames	140.00	Ton
Seals	190.00	Ton
Expanded metal lathe	0.08	SF
Closure plates	95.00	Ton
Instrument connection	200.00	Boiler

5.6 Recovery boiler installation estimate

5.6.1 Erect drum, generating bank, integral economizers multiple header type sheet 1

	Installation man-hour				
	Historical			Estimate	
	MH	Unit	Qty	Unit	BM
Erect drum, generating bank, and integral economizer					<u>0</u>
Install steam drum					
Steam drum (200,001 LB <= weight <= 300,000 LB)	3.80	Ton	0	Ton	0
Steam drum trim piping	260.00	Boiler	0	Boiler	0
Drum internals—remove and reinstall	136.00	Boiler	0	Boiler	0
Generating bank					
Generating tube sections and headers—unload and handle	10.00	Ton	0	Ton	0
Set rigging for hoisting assemblies	60.00	Boiler	0	Boiler	0
Hoist into place	1.50	Ton	0	Ton	0
Weld tubes to header stubs—2-1/2″ diameter, TIG no PWHT	2.40	EA	0	EA	0
Fit and weld filler bar at tube welds	1.00	Space	0	Space	0
Integral economizers multiple header type					
Economizer tube sections and headers—unload and handle	10.00	Ton	0	Ton	0
Set rigging for hoisting assemblies	60.00	Boiler	0	Boiler	0
Hoist into place	1.50	Ton	0	Ton	0
Weld tubes to header stubs—2-1/2″ diameter, TIG no PWHT	2.40	EA	0	EA	0
Fit and weld filler bar at tube welds	1.00	Space	0	Space	0
Baffles—place and weld including clips	60.00	Ton	0	Ton	0
Vibration bars	65.00	Ton	0	Ton	0
Buckstays	40.00	Ton	0	Ton	0
Casing and hoppers	55.00	Ton	0	Ton	0
Doors	55.00	Ton	0	Ton	0

5.6.2   Erect headers/panels, wall/roof tubes, weld tubes and panels sheet 2

	Historical			Installation man-hour Estimate	
	MH	Unit	Qty	Unit	BM
Erect headers/panels/wall/roof and weld tubes and panels					<u>0</u>
Boiler sidewall tubes					
Wall panels—shop assembled with headers	10.00	Ton	0	Ton	0
Fit and weld filler joining membrane panels	0.70	LF	0	LF	0
Furnace front wall tubes					
Headers—loose	15.00	Ton	0	Ton	0
Wall panels—shop assembled without headers	10.00	Ton	0	Ton	0
Weld tubes to header stubs—2-1/2" diameter, TIG no PWHT	2.40	EA	0	EA	0
Fit and weld filler bar at tube welds	1.00	Space	0	Space	0
Fit and weld filler joining membrane panels	0.70	LF	0	LF	0
Furnace roof tubes					
Headers—loose	30.00	Ton	0	Ton	0
Tubes—loose, furnace including screen	30.00	Ton	0	Ton	0
Weld tubes to header stubs—3" diameter, TIG no PWHT	3.00	EA	0	EA	0
Furnace sidewall tubes					
Headers—loose	15.00	Ton	0	Ton	0
Wall panels—shop assembled without headers	10.00	Ton	0	Ton	0
Weld tubes to header stubs—2-1/2" diameter, TIG no PWHT	2.40	EA	0	EA	0
Fit and weld filler bar at tube welds	1.00	Space	0	Space	0
Fit and weld filler joining membrane panels	0.70	LF	0	LF	0
Furnace rear wall tubes					
Furnace rear wall tubes (wall panel—membrane)	12.00	Ton	0	Ton	0
Furnace rear wall tubes (loose)	30.00	Ton	0	Ton	0
Weld tubes to header stubs—2-1/2" diameter, TIG no PWHT	2.40	EA	0	EA	0
Fit and weld filler bar at tube welds	1.00	Space	0	Space	0
Fit and weld filler joining membrane panels	0.70	LF	0	LF	0

5.6.3   Erect super heater, supply/riser down comers, attemperators sheet 3

			Installation man-hour		
	Historical			Estimate	
	MH	Unit	Qty	Unit	BM
Superheater, supply/riser/downcomers and attemperators					**0**
Superheaters					
Tubes—connecting	32.00	Ton	0	Ton	0
Tubes—sections	10.00	Ton	0	Ton	0
Headers	15.00	Ton	0	Ton	0
Weld superheater tubes; 1.77″ OD × 0.150″ WT, BW (TIG)	2.10	EA	0	EA	0
Attemperators for recovery units					
Spray type—complete with body and internals (no welding)	20.00	Ton	0	Ton	0
Drum type—internals, piping, valves, and supports (no welding)	45.00	Ton	0	Ton	0
Supply and riser tubes					
Headers	15.00	Ton	0	Ton	0
Tubes—make up or supply	25.00	Ton	0	Ton	0
Tubes—riser	29.00	Ton	0	Ton	0
Weld tubes to header stubs—4-1/2″ diameter, TIG no PWHT	4.00	EA	0	EA	0
Downcomers					
Install downcomer	10.00	Ton	0	Ton	0
Downcomers; 10″ OD × 0.500 WT BW	0.55	DI	0	DI	0
Downcomers; 12″ OD × 0.562 WT BW	1.05	DI	0	DI	0
Downcomers; 14″ OD × 0.594 WT BW	1.05	DI	0	DI	0

5.6.4   Install spout and hood and auxiliary boiler steel and equipment, sheet 4

				Installation man-hour		
	Historical			Estimate		
	MH	Unit	Qty	Unit	BM
Smelt spout and hood	55.00	Ton	0	Ton	**0**
Auxiliary boiler support steel and equipment					**0**
Floor and hopper support steel	28.00	Ton	0	Ton	0
Suspension steel	30.00	Ton	0	Ton	0
Buckstays	36.00	Ton	0	Ton	0
Casing	95.00	Ton	0	Ton	0
Burner wind box—primary, secondary, and tertiary	75.00	Ton	0	Ton	0
Primary burners	72.00	EA	0	EA	0
Lighters	5.00	EA	0	EA	0
Hoppers	95.00	Ton	0	Ton	0
Doors and frames	140.00	Ton	0	Ton	0
Seals	190.00	Ton	0	Ton	0
Expanded metal lathe	0.08	SF	0	SF	0
Closure plates	95.00	Ton	0	Ton	0
Instrument connection	200.00	Boiler	0	Boiler	**0**

5.7   Recovery boiler installation man-hours

	Estimated	Actual
	BM	BM
Erect drum, generating bank, and integral economizer	0	0
Erect headers/panels/wall/roof and weld tubes and panels	0	0
Superheater, supply/riser/downcomers, and attemperators	0	0
Smelt spout and hood	0	0
Auxiliary boiler support steel and equipment	0	0
Instrument connection	0	0
Recovery boiler installation man-hours	0	0

5.8 General scope of field work required for each reheat recovery boiler sheet 1

5.8.1 Scope of work-field erection

Pressure parts

Install steam drum
Drums include straps and/or U-bolts: steam drum
Steam drum trim piping
Drum internals—remove and reinstall
Generating bank
Generating tube sections and headers—unload and handle
Set rigging for hoisting assemblies
Hoist into place
Weld tubes to header stubs—2-1/2″ diameter, TIG weld
Fit and weld filler bar at tube welds
Integral economizers for recovery units—multiple header type
Economizer tube sections and headers—unload and handle
Set rigging for hoisting assemblies
Hoist into place
Weld tubes to header stubs—2-1/2″ diameter, TIG weld
Fit and weld filler bar at tube welds
Baffles—place and weld including clips
Vibration bars
Buckstays
Casing and hoppers
Doors
Headers, furnace, and waterwall panels
Boiler sidewall tubes
Wall panels—shop assembled with headers
Fit and weld filler joining membrane panels
Furnace front wall tubes
Headers—loose
Wall panels—shop assembled without headers
Weld tubes to header stubs—2-1/2″ diameter, TIG weld
Fit and weld filler bar at tube welds
Fit and weld filler joining membrane panels
Furnace roof tubes
Headers—loose
Tubes—loose, furnace including screen
Weld tubes to header stubs—3″ diameter, TIG weld
Furnace sidewall tubes
Headers—loose
Wall panels—shop assembled without headers
Weld tubes to header stubs—2-1/2″ diameter, TIG weld
Fit and weld filler bar at tube welds
Fit and weld filler joining membrane panels

5.9   General scope of field work required for each reheat recovery boiler sheet 2

5.9.1   Scope of work-field erection

Pressure parts

Furnace rear wall tubes
Furnace rear wall tubes (wall panel—membrane)
Furnace rear wall tubes (loose)
Weld tubes to header stubs—2-1/2″ diameter, TIG weld
Fit and weld filler bar at tube welds
Fit and weld filler joining membrane panels
Superheaters for reheat recovery boiler
Tubes—connecting
Tubes—sections
Headers
Weld superheater tubes; 1.77″ OD × 0.150″ WT, BW (TIG)
Attemperators for recovery units
Spray type—complete with body and internals (excludes welding)
Drum type—complete with internals, piping, valves, and supports (excludes welding)
Reheater for reheat recovery boiler
Tubes—connecting
Tubes—sections
Headers
Weld reheater tubes; 2″ OD, BW (TIG)
Supply and riser tubes
Headers
Tubes—make up or supply
Tubes—riser
Weld tubes to header stubs—4-1/2″ diameter, TIG weld
Downcomers
Install downcomer
Diameter weld, girth
Diameter weld, ring
Diameter weld, ring
Smelt spout and hood
Auxiliary equipment
Floor and hopper support steel
Suspension steel
Buckstays
Casing
Burner wind box—primary, secondary, and tertiary
Primary burners

Lighters
Hoppers
Doors and frames
Seals
Expanded metal lathe
Closure plates
Instrument connection

5.10   General scope of field work required for each reheat recovery boiler sheet 3

5.10.1   Scope of work-field erection

Auxiliary equipment

Tubular air heater
Tube sheets
Baffle sheets—turn and vibration
Supports—tube sheets
Erect grid support steel
Supports—main
Tubes—steel
Expand 2″ ends
Expand 2-1/2″ ends
Casing, veneer including batten bars and channels
Casing, main and hoppers including expansion joints and doors
Dampers
Flues and ducts for reheat recovery boiler
Flues and ducts—including supports, hoppers, doors, expansion joints, and dampers
Coupling guards

5.11   Reheat recovery boiler estimate data

5.11.1   Erect drum, generating bank, integral economizers multiple header type sheet 1

Description	MH	Unit
Drums include straps and/or U-bolts		
100,001–200,000 lbs	4.60	Ton
200,001–300,000 lbs	3.80	Ton
Steam drum trim piping	260.00	LOT
Drum internals—remove and reinstall	136.00	Boiler
Generating bank		
Generating tube sections and headers—unload and handle	10.00	Ton
Set rigging for hoisting assemblies	60.00	Boiler
Hoist into place	1.50	Ton
Weld tubes to header stubs—2-1/2″ diameter, TIG No PWHT	2.60	EA
Fit and weld filler bar at tube welds	1.00	Space
Integral economizers multiple header type		
Economizer tube sections and headers—unload and handle	10.00	Ton
Set rigging for hoisting assemblies	60.00	Boiler
Hoist into place	1.50	Ton
Weld tubes to header stubs—2-1/2″ diameter, TIG No PWHT	2.60	EA
Fit and weld filler bar at tube welds	1.00	Space
Baffles—place and weld including clips	60.00	Ton
Vibration bars	65.00	Ton
Buckstays	40.00	Ton
Casing and hoppers	55.00	Ton
Doors	55.00	Ton

5.11.2   Erect headers and panels, furnace wall and roof tubes, weld tubes and panels sheet 2

Description	MH	Unit
Erect headers and wall panels, fit and weld waterwall panels		
Boiler sidewall, furnace front wall, roof, sidewall and rear wall tubes		
Wall panels—shop assembled with headers	10.00	Ton
Fit and weld filler joining membrane panels	0.70	LF
Headers—loose	15.00	Ton
Wall panels—shop assembled without headers	10.00	Ton
Weld tubes to header stubs—2-1/2″ diameter, TIG No PWHT	2.60	EA
Headers—loose	30.00	Ton
Tubes—loose, furnace including screen	30.00	Ton
Weld tubes to header stubs—3″ diameter, TIG No PWHT	4.00	EA
Furnace rear wall tubes (wall panel—membrane)	12.00	Ton
Furnace rear wall tubes (loose)	30.00	Ton
Fit and weld filler bar at tube welds	1.00	Space
Fit and weld filler joining membrane panels	0.70	LF

5.11.3   Erect/weld SH, reheater, supply/riser tube, down comers, attemperators, sheet 3

Description	MH	Unit
Superheater, reheater, supply and riser tubes, and downcomers		
Tubes—connecting	32.00	Ton
Tubes—sections	10.00	Ton
Headers	15.00	Ton
Weld superheater tubes; 1.77″ OD × 0.150″ WT, BW (TIG)	2.10	EA
Weld reheater tubes; 2″ OD, BW (TIG)	2.30	EA
Attemperators for recovery units		
Spray type—complete with body and internals (excludes welding)	20.00	Ton
Drum type—complete with internals, piping, valves, and supports (excludes welding)	45.00	Ton
Headers	15.00	Ton
Tubes—make up or supply	25.00	Ton
Tubes—riser	29.00	Ton
Weld tubes to header stubs—4-1/2″ diameter, TIG No PWHT	4.40	EA
Install downcomer	10	Ton
Downcomers; 10″ OD × 0.500 WT BW	0.55	DI
Downcomers; 12″ OD × 0.562 WT BW	1.05	DI

5.11.4   Install smelt spout and hood and auxiliary boiler support steel and equipment, sheet 4

Description	MH	Unit
Smelt spout and hood	55.00	Ton
Auxiliary boiler support steel and equipment		
Floor and hopper support steel	28.00	Ton
Suspension steel	30.00	Ton
Buckstays	36.00	Ton
Casing	95.00	Ton
Burner wind box—primary, secondary, and tertiary	75.00	Ton
Primary burners	72.00	EA
Lighters	5.00	EA
Hoppers	95.00	Ton
Doors and frames	140.00	Ton
Seals	190.00	Ton
Expanded metal lathe	0.08	SF
Closure plates	95.00	Ton
Instrument connection	200.00	Boiler

5.11.5   Install tubular air heater, flues and ducts

Description	MH	Unit
Tubular air heater		
Tube sheets	28.00	Ton
Baffle sheets—turn and vibration	60.00	Ton
Supports—tube sheets	20.00	Ton
Erect grid support steel	28.00	Ton
Supports—main	18.00	Ton
Tubes—steel	9.00	Ton
Expand 2″ ends	0.65	EA
Expand 2-1/2″ ends	0.95	EA
Casing, veneer including batten bars and channels	64.00	Ton
Casing, main and hoppers including expansion joints and doors	60.00	Ton
Dampers	40.00	Ton
Flues and ducts		
Flues and ducts—supports, hoppers, doors, expansion joints, and dampers	40.00	Ton
Coupling guards	14.00	Ton

5.12   Reheat recovery boiler installation estimate

5.12.1   Erect drum, generating bank, integral economizers multiple header type sheet 1

	Installation man-hour				
	Historical			Estimate	
	MH	Unit	Qty	Unit	BM
Erect steam drum, install generating bank and weld tubes, install integral economizers					<u>0</u>
Install steam drum					
Steam drum (200,001 LB <=weight<=300,000 LB)	3.80	Ton	0	Ton	0
Steam drum trim piping	260.00	Boiler	0	Boiler	0
Drum internals—remove and reinstall	136.00	Ton	0	Ton	0
Generating bank					
Generating tube sections and headers—unload and handle	10.00	Ton	0	Ton	0
Set rigging for hoisting assemblies	60.00	Boiler	0	Boiler	0
Hoist into place	1.50	Ton	0	Ton	0
Weld tubes to header stubs—2-1/2″ diameter, TIG no PWHT	2.60	EA	0	EA	0
Fit and weld filler bar at tube welds	1.00	Space	0	Space	0
Integral economizers multiple header type					
Economizer tube sections and headers—unload and handle	10.00	Ton	0	Ton	0
Set rigging for hoisting assemblies	60.00	Boiler	0	Boiler	0
Hoist into place	1.50	Ton	0	Ton	0
Weld tubes to header stubs—2-1/2″ diameter, TIG no PWHT	2.60	EA	0	EA	0
Fit and weld filler bar at tube welds	1.00	Space	0	Space	0
Baffles—place and weld including clips	60.00	Ton	0	Ton	0
Vibration bars	65.00	Ton	0	Ton	0
Backstays	40.00	Ton	0	Ton	0
Casing and hoppers	55.00	Ton	0	Ton	0
Doors	55.00	Ton	0	Ton	0

5.12.2   Erect headers, panels, furnace wall, roof tubes, weld tubes and panels sheet 2

	Historical		Estimate		
	MH	Unit	Qty	Unit	BM
Erect headers, panels, furnace wall, roof tubes					<u>0</u>
Boiler sidewall tubes					
Wall panels—shop assembled with headers	10.00	Ton	0	Ton	0
Fit and weld filler joining membrane panels	0.70	LF	0	LF	0
Furnace front wall tubes					
Headers—loose	15.00	Ton	0	Ton	0
Wall panels—shop assembled without headers	10.00	Ton	0	Ton	0
Weld tubes to header stubs—2-1/2″ diameter, TIG no PWHT	2.60	EA	0	EA	0
Fit and weld filler bar at tube welds	1.00	Space	0	Space	0
Fit and weld filler joining membrane panels	0.70	LF	0	LF	0
Furnace roof tubes					
Headers—loose	30.00	Ton	0	Ton	0
Tubes—loose, furnace including screen	30.00	Ton	0	Ton	0
Weld tubes to header stubs—3″ diameter, TIG no PWHT	4.00	EA	0	EA	0
Furnace sidewall tubes					
Headers—loose	15.00	Ton	0	Ton	0
Wall panels—shop assembled without headers	10.00	Ton	0	Ton	0
Weld tubes to header stubs—2-1/2″ diameter, TIG no PWHT	2.60	EA	0	EA	0
Fit and weld filler bar at tube welds	1.00	Space	0	Space	0
Fit and weld filler joining membrane panels	0.70	LF	0	LF	0
Furnace rear wall tubes					
Furnace rear wall tubes (wall panel—membrane)	12.00	Ton	0	Ton	0
Furnace rear wall tubes (loose)	30.00	Ton	0	Ton	0
Weld tubes to header stubs—2-1/2″ diameter, TIG no PWHT	2.60	EA	0	EA	0
Fit and weld filler bar at tube welds	1.00	Space	0	Space	0
Fit and weld filler joining membrane panels	0.70	LF	0	LF	0

5.12.3   Erect superheater, reheater, riser tube, down comers, attemperators, sheet 3

	Installation man-hour				
	Historical			Estimate	
	MH	Unit	Qty	Unit	BM
Erect, superheater, reheater, tubes, downcomers, attemperators					<u>0</u>
Superheaters					
Tubes—connecting	32.00	Ton	0	Ton	0
Tubes—sections	10.00	Ton	0	Ton	0
Headers	15.00	Ton	0	Ton	0
Weld superheater tubes; 1.77″ OD × 0.150″ WT, BW (TIG)	2.10	EA	0	EA	0
Attemperators for recovery units					
Spray type—complete with body and internals (excludes welding)	20.00	Ton	0	Ton	0
Drum type—complete with internals, piping, valves, and supports (excludes welding)	45.00	Ton	0	Ton	0
Reheater					
Tubes—connecting	32.00	Ton	0	Ton	0
Tubes—sections	10.00	Ton	0	Ton	0
Headers	15.00	Ton	0	Ton	0
Weld reheated tubes; 2″ OD, BW (TIG)	2.30	EA	0	Ton	0
Supply and riser tubes					
Headers	15.00	Ton	0	Ton	0
Tubes—make up or supply	25.00	Ton	0	Ton	0
Tubes—riser	29.00	Ton	0	Ton	0
Weld tubes to header stubs—4-1/2″ diameter, TIG no PWHT	4.40	EA	0	EA	0
Downcomers					
Install downcomer	10.00	Ton	0	Ton	0
Downcomers; 10″ OD × 0.500 WT BW	0.55	DI	0	DI	0
Downcomers; 12″ OD × 0.562 WT BW	1.05	DI	0	DI	0
Downcomers; 14″ OD × 0.594 WT BW	1.05	DI	0	DI	0

5.12.4 Install smelt spout and hood and auxiliary boiler support steel and equipment, sheet 4

	Installation man-hour				
	Historical			Estimate	
	MH	Unit	Qty	Unit	BM
Smelt spout and hood	55.00	Ton	0	Ton	0
Auxiliary boiler support steel and equipment					0
Floor and hopper support steel	28	Ton	0	Ton	0
Suspension steel	30	Ton	0	Ton	0
Buckstays	36	Ton	0	Ton	0
Casing	95	Ton	0	Ton	0
Burner wind box—primary, secondary, and tertiary	75	Ton	0	Ton	0
Primary burners	72.00	EA	0	EA	0
Lighters	5.00	EA	0	EA	0
Hoppers	95	Ton	0	Ton	0
Doors and frames	140	Ton	0	Ton	0
Seals	190	Ton	0	Ton	0
Expanded metal lathe	0.08	SF	0	SF	0
Closure plates	95	Ton	0	Ton	0
Instrument connection	200	Boiler	0	Boiler	0

5.12.5   Install tubular air heater, flues and ducts

	Installation man-hour				
	Historical			Estimate	
	MH	Unit	Qty	Unit	BM
Tubular air heater					**0**
Tube sheets	28.00	Ton	0	Ton	0
Baffle sheets—turn and vibration	60.00	Ton	0	Ton	0
Supports—tube sheets	20.00	Ton	0	Ton	0
Erect grid support steel	28.00	Ton	0	Ton	0
Supports—main	18.00	Ton	0	Ton	0
Tubes—steel	9.00	Ton	0	Ton	0
Expand 2″ ends	0.65	EA	0	EA	0
Expand 2-1/2″ ends	0.95	EA	0	EA	0
Casing, veneer including batten bars and channels	64.00	Ton	0	Ton	0
Casing, main and hoppers including expansion joints and doors	60.00	Ton	0	Ton	0
Dampers	40.00	Ton	0	Ton	0
Flues and ducts					**0**
Flues and ducts—supports, hoppers, doors, expansion joints, and dampers	40.00	Ton	0	Ton	0
Coupling guards	14.00	Ton	0	Ton	0

5.13   Reheat recovery boiler installation man-hours

	Estimated BM	Actual BM
Erect steam drum, install generating bank and weld tubes, install integral economizers	0	0
Erect headers and panels, furnace wall and roof tubes, weld tubes and panels	0	0
Erect, weld superheater, reheater, supply, riser tubes, downcomers, attemperators	0	0
Smelt spout and hood	0	0
Auxiliary boiler support steel and equipment	0	0
Instrument connection	0	0
Tubular air heater	0	0
Flues and ducts	0	0
Reheat recovery boiler installation man-hours	0	0

Chapter 6

Industrial package boiler

6.1 Industrial package boilers

A-type, D-type, and O-type package boilers are all watertube boilers.

A-type package boiler has two water drums, one steam drum, and generating tubes. A-type package boilers are smaller than D-type and will fit smaller plants, but do not have the same power output as D-type package boilers.

O-type package boiler is smaller compared with D-type and A-type package boilers. O-type package boilers consist of one water drum and one steam drum. Generating tubes are lined up from either sides of the steam and water drums. O-type boilers are mainly used for fast steam production and reduced maintenance cost.

D-type package boiler has water drum, steam drum, and generating tubes. Shipped shaped into a D-shape, hence the name D-type package boiler. Fully shop assembled (packaged), two-drum, bottom-supported, which can be shipped by rail, truck, or barge.

Pressure 150–2000 PSIG
Temperature saturated and superheat up to 1050°F
Fuel: natural gas, various waste, and off-gasses, #2-#6 oil
Steam capacities up to 300,000 lb/h

6.1.1 Modular design

A two-drum, bottom-supported multiple gas pass unit, prefabricated modular construction for field assembly.

Capacity: 100,000–700,000 lb/h
Steam pressure: to 1150 PSIG
Steam temperature: to 960°F
Fuels: liquid and/or gaseous fuels, blast furnace gas, coke oven gas, and various combustible by-product gases and liquids

Industrial Process Plant Construction Estimating and Man-Hour Analysis.
https://doi.org/10.1016/B978-0-12-818648-0.00006-5
101

6.2 General scope of field work required for each D-type package boiler sheet 1

6.2.1 Scope of work-field erection

Fully shop assembled (packaged), two-drum, bottom-supported, which can be shipped by rail, truck, or barge.
Unload, place, and set package boiler on foundation
Install boiler piping
Flues and ducts
Erect stack
Stack—knocked down
Stack—shop assembled
Auxiliary equipment
Unload, place, and set economizer (shop assembled)
Platform steel
Support steel
Grating, stairs, and handrails
Feed water regulators

6.3 D-type package boiler estimate data

6.3.1 Set boiler and install piping, flues, stack, economizer, and structural steel sheet 1

Description	MH	Unit
Unload, place, and set package boiler on foundation	8.0	Ton
Install boiler piping	140.0	Unit
Stack—knocked down (diameter <= 72″)	1.8	LF
Stack—shop assembled	28.0	Ton
Unload, place, and set economizer (shop assembled)	20.0	Ton
Structural steel	24.0	Ton
Platform framing	0.2	SF
Grating	0.2	SF
Handrail	0.3	LF
Stair treads	0.9	EA
Feed water regulators	55.0	EA

6.4   D-type package boiler installation estimate

6.4.1   Boiler and piping, flues, stack, economizer, and structural steel sheet 1

	Historical			Estimate	
	MH	Unit	Qty	Unit	BM
Boiler and piping, flues, stack, economizer, and structural sheet 1					<u>0</u>
Unload, place, and set package boiler on foundation	8.0	Ton	0	Ton	0
Install boiler piping	140.0	Unit	0	Unit	0
Erect stack					
Stack—knocked down	1.8	LF	0	LF	0
Stack—shop assembled	28.0	Ton	0	Ton	0
Unload, place, and set economizer (shop assembled)	20.0	Ton	0	Ton	0
Structural steel	24.0	Ton	0	Ton	0
Platform framing	0.2	SF	0	SF	0
Grating	0.2	SF	0	SF	0
Handrail	0.3	LF	0	LF	0
Stair treads	0.9	EA	0	EA	0
Feedwater regulators	55.0	EA	0	EA	0

6.4.2   D-type package boiler installation man-hours

	Estimated	Actual
	BM	BM
Boiler and piping, flues, stack, economizer, and structural sheet 1	0	0
D-type package boiler installation man-hours	<u>0</u>	<u>0</u>

6.5 General scope of field work required for each field erected package boiler sheet 1

6.5.1 Scope of work-field erection

A two-drum, bottom-supported multiple gas pass unit, prefabricated, knocked down for field assembly and erection.

Pressure parts

Install steam and mud drums; include straps or U-bolts and miscellaneous welding
Steam drum trim piping
Drum internals—remove and reinstall
Generating bank tubes
Install generating tubes
2-1/2″ OD × 0.203″ wall thickness, swage and roll
Baffles, metallic; place and weld with clips
Floor and hopper support steel; place and weld
Tube furnace and convection
Outer casing includes welding
Inner casing includes welding
Wind box
Burner
Weight <= 600 lb
600 lb < weight <= 1000 lb
Weight > 1000 lb
Floor plate
Sootblower wall boxes include welding
Sootblower piping includes welding
Pipe and fittings
Doors and frames; place and weld to tubes and/or casing
Feedwater regulator
Shop-assembled superheaters
Tube welding; over 2-1/2″ and including 3″ TIG (1000 PSI)
Auxiliary equipment
Air heater
Support steel and interconnecting ductwork
Damper
Damper and junction piece
Set and assemble forced draft (FD) fan

6.6   Field erected package boiler estimate data

6.6.1   Drums, tubes baffles, steel, casing, wind box, burner, sheet 1

Description	MH	Unit
Drums include straps and/or U-bolts		
Steam drum and drum saddles	10.0	Ton
Mud drum and drum saddles	10.0	Ton
Steam drum trim piping	260.0	LOT
Drum internals—remove and reinstall	140.0	Boiler
Generating bank tubes		
Install generating tubes	0.5	EA
2'1/2″ ends—expand tubes in steam and mud drums	0.4	End
Baffles, metallic; place and weld with clips	16.0	EA
Floor and hopper support steel; place and weld	24.0	Ton
Tube furnace and convection	1.0	EA
Outer casing includes welding	1.4	LF
Inner casing includes welding	1.6	LF
Burner and wind box	59.40	Ton
Floor plate	0.30	SF
Sootblower wall Boxes include welding	10.0	EA
Sootblower piping includes welding	60	EA
Piping	180	Boiler
Doors and frames; place and weld to tubes and/or casing	10	EA
Feedwater regulator	55	EA
Shop-assembled superheaters	24	EA
Tube welding; over 2-1/2″ and including 3″ TIG (1000 PSI)	2.7	EA

6.7 Field erected package boiler installation estimate

6.7.1 Drums, tubes baffles, steel, casing, wind box, burner, sheet 1

	Historical			Estimate	
	MH	Unit	Qty	Unit	BM
Install steam and mud drums					**0**
Steam drum and drum saddles	10.0	Ton	0	Ton	0
Mud drum and drum saddles	10.0	Ton	0	Ton	0
Steam drum trim piping	0.0	LOT	0	LOT	0
Drum internals—remove and reinstall	140.0	Boiler	0	Boiler	0
Generating bank tubes					
2-1/2″ OD × 0.203″ wall thickness, swage, and roll					**0**
Install generating tubes	0.50	EA	0	EA	0
2′1/2″ ends—expand tubes in steam and mud drums	0.4	End	0	End	0
Baffles, metallic; place and weld with clips	16.0	EA	0	EA	**0**
Floor and hopper support steel; place and weld	24.0	Ton	0	Ton	**0**
Tube furnace and convection	1.0	EA	0	EA	**0**
Burner and wind box	59.40	Ton	0	Ton	**0**
Floor plate	0.30	SF	0	SF	**0**
Sootblower wall boxes include welding	10.0	EA	0	EA	**0**
Sootblower piping includes welding	60	EA	0	EA	**0**
Piping	180	Boiler	0	Boiler	**0**
Doors and frames; place and weld to tubes and/or casing	10	EA	0	EA	**0**
Feedwater regulator	55	EA	0	EA	**0**
Shop-assembled superheaters	24	EA	0	EA	**0**
Tube welding; over 2-1/2″ and including 3″ TIG (1000 PSI)	2.7	EA	0	EA	**0**

6.8 Field erected package boiler installation man-hours estimated actual

	BM	BM
Install steam and mud drums	0	0
Generating bank tubes	0	0
Baffles, metallic; place and weld with clips	0	0
Floor and hopper support steel; place and weld	0	0
Tube furnace and convection	0	0
Burner and wind box	0	0
Floor plate	0	0
Sootblower wall boxes include welding	0	0
Sootblower piping includes welding	0	0
Piping	0	0
Doors and frames; place and weld to tubes and/or casing	0	0
Feedwater regulator	0	0
Shop-assembled superheaters	0	0
Tube welding; over 2-1/2″ and including 3″ TIG (1000 PSI)	0	0
Field-erected package boiler installation man-hours	0	0

6.9 General scope of field work required for each air heater

Erect support steel
Set air heater modules
Set plenums
Erect interconnecting ductwork

6.10   Air heater estimate data

6.10.1   Air heater sheet 1

Description	MH	Unit
Support steel		
W 8 × 31 × 30' × 3EA	24.00	Ton
Structural; legs with base plates	4.00	EA
Shim/set baseplates	2.00	EA
Make up baseplates	2.60	EA
Grout baseplates	4.00	EA
W 8 × 31 × 42' × 3 EA horizontal beam	24.00	Ton
Bolted connection	1.50	EA
Tubular air heater modules	2.70	Ton
Plenum	11.00	Ton
Erect interconnecting ductwork		
Duct transition, reducer, elbow, and sections	32.00	Ton
Bolted connection	0.26	EA

6.11   Air heater installation estimate

6.11.1   Air heater sheet 1

	Historical			Estimate	
	MH	Unit	Qty	Unit	BM
Support steel					**0**
W 8 × 31 × 30' × 3EA	24.00	Ton	0	Ton	0
Structural; legs with baseplates	4.00	EA	0	EA	0
Shim/set baseplates	2.00	EA	0	EA	0
Make up baseplates	2.60	EA	0	EA	0
Grout baseplates	4.00	EA	0	EA	0
W 8 × 31 × 42' × 3 EA horizontal beam	24.00	Ton	0	Ton	0
Bolted connection	1.50	EA	0	EA	0
Air heater modules	2.70	Ton	0	Ton	0
Plenum	11.00	Ton	0	Ton	0
Erect interconnecting ductwork					**0**
Duct transition	32.00	Ton	0	Ton	0
Bolted connection-duct transition to air heater module	0.26	EA	0	EA	0
Duct sections	32.00	Ton	0	Ton	0
Duct reducer	32.00	Ton	0	Ton	0
Duct elbow	32.00	Ton	0	Ton	0
Bolted connection-duct sections	0.26	EA	0	EA	0

6.12   Air heater installation man-hours

	Estimated BM	Actual BM
Support steel	0	0
Tubular air heater modules	0	0
Plenum	0	0
Erect interconnecting ductwork	0	0
Air heater installation man-hours	**0**	**0**

6.13   General scope of field work required for FD fan

6.13.1   Fans

Perform all millwright work for the four fans. Alignment and grouting.

FD fan

Install FD fan
Work scope:
Set housing and inlet box
Set shaft and inlet cones
Install motor sole plates
Set inlet damper and damper drive
Set motor
Set inlet silencer
Lubrication
Laser alignment
Final dowelling
Grouting of bearings and motor

6.14   FD fan estimate data
6.14.1   FD fan sheet 1

Description	MH	Unit
Fans		
Install sole and baseplates	2.0	EA
Attach fan to foundation-bolted connection	0.4	EA
FD fan		
Set fan housing split	6.0	EA
Set impeller assembly/shaft	8.0	EA
Alignment fan bearings	16.0	EA
Seal weld housing inside/outside	20.0	JOB
Weld inlet cones and stud rings	6.0	JOB
Weld shaft seal stud rings	6.0	JOB
Coupling alignment	14.0	JOB
Install damper	16.0	EA
Steam coil air heater	28.5	Ton

6.15   FD fan installation estimate
6.15.1   FD fan sheet 1

	Historical			Estimate	
	MH	Unit	Qty	BM	MW
Set and assemble FD fan				**0**	**0**
Install sole and baseplates	2.0	EA	0		0
Set fan housing split	6.0	EA	0	0	
Set impeller assembly/shaft	8.0	EA	0	0	
Alignment fan bearings	16.0	EA	0		0
Seal weld housing inside/outside	20.0	JOB	0	0	
Weld inlet cones and stud rings	6.0	JOB	0	0	
Weld shaft seal stud rings	6.0	JOB	0	0	
Coupling alignment	14.0	JOB	0		0
Install damper	16.0	EA	0	0	
Attach fan to foundation-bolted connection	0.4	EA	0	0	
Steam coil air heater	28.5	Ton	**0**	**0**	

6.16   FD Fan Installation Man-Hours

	Estimated		Actual	
	BM	MW	MH	
Set and assemble FD fan	0	0	0	0
Steam coil air heater	0		0	0
FD fan installation man-hours	0	0	0	0

6.17   Field erected package boiler man-hour breakdown

	Estimate		
	BM	IW	MH
Field-erected package boiler installation	**0**		**0**
Air heater installation	**0**		**0**
FD fan installation	**0**	**0**	**0**
Field-erected package boiler man-hours	**0**	**0**	**0**

Chapter 7

Diesel power plant

7.1 Section introduction—Piping and equipment

This section provides the reader a basic understanding of the fundamentals and the operating relationship between the plant equipment in a diesel power plant and covers the craft labor for the assembly and field erection required to put the equipment into operation in a diesel power plant.

7.1.1 Diesel power plant

The diesel power plant is an electric installation equipped with one or several electric current generators driven by diesel engines. Stationary diesel power plants use four-stroke diesel engines. Diesel power plants are classed as average in their power rating if the rating does not exceed 750 KW; large diesel power plants can have a power rating of 2200 kW or more. The advantages of a diesel power plant are favorable economy of operation, stable operation characteristics, and an easy and quick start-up.

Industrial Process Plant Construction Estimating and Man-Hour Analysis.
https://doi.org/10.1016/B978-0-12-818648-0.00007-7
113

7.2 General scope of field work required for diesel power plant

7.2.1 Scope of work-field erection

Engines, couplings, and generators (B2)
Engine and auxiliary platforms (A3)
Pipe modules (B3)
Exhaust gas system (A11)
Selective catalytic reduction (SCR) and urea system (A13)
Cooling system (A10)
Intake air system (A11)
Fire water system (A14)
Utility gas heating system
Lube oil system (pump and tanks)
Utility water
Secondary pipe support, material from water for injection (WFI)
Valves list
Manometer/thermometer list

7.3 Engines, couplings, and generators (B2) estimate data

Description	MH	Unit
Engines, couplings, and generators (B2)		
Engine generator set 296962#	199.4	EA
Spring element	3.3	EA
Anchoring plate with anchor bolt (AB)	3.3	EA

7.4 Engines, couplings, and generators (B2) estimate

Description	MH	Qty	Unit	Qty	Unit	BM	IW
Engines, couplings, and generators (B2)						**0**	**0**
Engine generator set 296962#	199.4	14	EA	0	EA		0
Spring element	3.3	280	EA	0	EA		0
Anchoring plate with AB	3.3	280	EA	0	EA		0

7.5 Engines, couplings, and generators (B2) installation man hours

Facility—diesel power plant	Actual	Estimate			
Description	MH	BM	IW	MW	MH
Engine generator set 296962#	2792	0	0	0	0
Spring element	924	0	0	0	0
Anchoring plate with AB	924	0	0	0	0
Equipment installation man-hours	4640	0	0	0	0

7.6 Engines and Aux platforms (A3) estimate data

Description	MH	Unit
Engine and auxiliary platforms (A3)		
Extension platform legs	1.7	EA
Handrail	0.6	LF
Sleeve single	0.1	EA
Sleeve double	1.4	EA
Platform frame	4.8	EA
Platform legs	2.7	EA
Stair	4.5	EA
Stair railing	2.3	LF
Grating	1.0	SF
Ventilation unit (engine hall/aux area)	10.0	EA
Engine auxiliary module	44.3	EA
Stair with handrail	22.9	EA
20″ Charge air pipe	4.3	LF
20″ Field weld	20.0	EA
20″ Flange weld	19.3	EA
20″ Bolt up	8.6	EA

7.7 Engines and Aux platforms (A3) estimate

Description	MH	Qty	Unit	Qty	Unit	BM	IW	PF
Engine and auxiliary platforms (A3)						<u>0</u>	<u>0</u>	<u>0</u>
Extension platform legs	1.7	168	EA	0	EA	0		
Handrail	0.6	672	LF	0	LF	0		
Sleeve single	0.1	1708	EA	0	EA	0		
Sleeve double	1.4	28	EA	0	EA	0		
Platform frame	4.8	84	EA	0	EA	0		
Platform legs	2.7	112	EA	0	EA	0		
Stair	4.5	84	EA	0	EA	0		
Stair railing	2.3	168	LF	0	LF	0		
Grating	1.0	138	SF	0	SF	0		
Ventilation unit (engine hall/ aux area)	10.0	32	EA	0	EA	0		
Engine auxiliary module	44.3	14	EA	0	EA		0	
Stair with handrail	22.9	14	EA	0	EA		0	
20″ Charge air pipe	4.3	28	LF	0	LF			0
20″ Field weld	20.0	14	EA	0	EA			0
20″ Flange weld	19.3	28	EA	0	EA			0
20″ Bolt up	8.6	28	EA	0	EA			0

7.8 Engines and Aux platforms (A3) installation man-hours

Facility—diesel power plant	Actual		Estimate		
Description	MH	BM	IW	PF	MH
Extension platform legs	280	0	0	0	0
Handrail	380	0	0	0	0
Sleeve single	180	0	0	0	0
Sleeve double	40	0	0	0	0
Platform frame	400	0	0	0	0
Platform legs	300	0	0	0	0
Stair	380	0	0	0	0
Stair railing	380	0	0	0	0
Grating	140	0	0	0	0
Ventilation unit (engine hall/aux area)	320	0	0	0	0
Engine auxiliary module	620	0	0	0	0
Stair with handrail	320	0	0	0	0
20″ Charge air pipe	120	0	0	0	0
20″ Field weld	280	0	0	0	0
20″ Flange weld	540	0	0	0	0
20″ Bolt up	240	0	0	0	0
Equipment installation man-hours	<u>4918</u>	<u>0</u>	<u>0</u>	<u>0</u>	<u>0</u>

7.9 Installation of pipe modules (B3) estimate data

Description	MH	Unit
Pipe module	11.5	EA
Bellows	5.4	EA
Bellows (bolt up)	3.8	EA
2.5" Rubber bellows (bolt up)	1.2	EA
IC pipe (1",1.5",2")	3.0	LF

7.10 Installation of pipe modules (B3) estimate

Description	MH	Qty	Unit	Qty	Unit	BM	IW	MW
Installation of pipe modules (B3) estimate						<u>0</u>	<u>0</u>	<u>0</u>
Pipe module	11.5	14	EA	0.0	EA	0		
Bellows	5.4	112	EA	0.0	EA	0		
Bellows (bolt up)	3.8	224	EA	0.0	EA	0		
2.5" Rubber bellows (bolt up)	1.2	56	EA	0.0	EA	0		
IC pipe (1",1.5",2")	3.0	84	LF	0.0	LF	0		

7.11 Installation of pipe modules (B3) installation man hours

Facility—diesel power plant	Actual	Estimate			
Description	MH	BM	IW	MW	MH
Pipe module	161	0	0	0	<u>0</u>
Bellows	600	0	0	0	<u>0</u>
Bellows (bolt up)	840	0	0	0	<u>0</u>
2.5" Rubber bellows (bolt up)	67	0	0	0	<u>0</u>
IC pipe (1",1.5",2")	252	0	0	0	<u>0</u>
Equipment installation man-hours	<u>1920</u>	<u>0</u>	<u>0</u>	<u>0</u>	<u>0</u>

7.12 Exhaust gas system (A11) estimate data

Description	MH	Unit
Exhaust gas system (A11)		
Exhaust gas module 11680#	33.0	EA
Bolt down	1.1	EA
Service platform for injection pipe	22.0	EA
Double bellows	14.8	EA
Double multi-ply bellows	14.8	EA
Exhaust gas pipe		
22″ Straight run	24.3	LF
22″ Bend	24.3	EA
22″ Extension pieces with explosion pipe flange	24.3	EA
22″ Field weld	23.8	EA
22″ Flange weld	23.3	EA
22″ Bolt up	9.0	EA
20″ Exhaust explosion pipe	27.5	LF
20″ Bolt up	10.1	EA
22″ Fixed/sliding support	10.0	EA
22″ Exhaust gas pipe (stack)	40.0	LF
22″ Flange weld	14.3	EA
22″ Bolt up	50.0	EA
20″ Rupture disk	10.0	EA
Support platform at explosion pipe		
Beams	2.0	EA
X-brace	4.1	EA
AB	2.5	EA
Caged ladder	10.0	EA
Platform	10.0	EA
Railing	10.0	LF
Expansion joint	10.0	EA

7.13 Exhaust gas system (A11) estimate

Description	MH	Qty	Unit	Qty	Unit	BM	IW	PF
Exhaust gas system (A11)						**0**	**0**	**0**
Exhaust gas module 11680#	33.0	14	EA	0	EA		0	
Bolt down	1.1	224	EA	0	EA		0	
Service platform for injection pipe	22.0	14	EA	0	EA		0	
Double bellows	14.8	16	EA	0	EA		0	
Double multi-ply bellows	14.8	3	EA	0	EA		0	
Exhaust gas pipe								
22″ Straight run	24.3	14	LF	0	LF			0
22″ Bend	24.3	14	EA	0	EA			0
22″ Extension pieces with explosion pipe flange	24.3	14	EA	0	EA			0
22″ Field weld	23.8	16	EA	0	EA			0
22″ Flange weld	23.3	48	EA	0	EA			0
22″ Bolt up	9.0	80	EA	0	EA			0
20″ Exhaust explosion pipe	27.5	8	LF	0	LF			0
20″ Bolt up	10.1	8	EA	0	EA			0
22″ Fixed/sliding support	10.0	28	EA	0	EA			0
22″ Exhaust gas pipe (stack)	40.0	14	LF	0	LF			0
22″ Flange weld	14.3	14	EA	0	EA			0
22″ Bolt up	50.0	14	EA	0	EA			0
20″ Rupture disk	10.0	8	EA	0	EA			0
Support platform at explosion pipe								
Beams	2.0	256	EA	0	EA		0	
X-brace	4.1	32	EA	0	EA		0	
AB	2.5	32	EA	0	EA		0	
Caged ladder	10.0	8	EA	0	EA		0	
Platform	10.0	8	EA	0	EA		0	
Railing	10.0	24	LF	0	LF		0	
Expansion joint	10.0	24	EA	0	EA		0	

7.14 Exhaust gas system (A11) installation man-hours

Facility—diesel power plant	Actual	Estimate			
Description	MH	BM	IW	PF	MH
Exhaust gas module 11680#	462	0	0	0	0
Bolt down	246	0	0	0	0
Service platform for injection pipe	308	0	0	0	0
Double bellows	236	0	0	0	0
Double multi-ply bellows	44	0	0	0	0
Exhaust gas pipe					
22″ Straight run	340	0	0	0	0
22″ Bend	340	0	0	0	0
22″ Extension pieces with explosion pipe flange	340	0	0	0	0
22″ Field weld	380	0	0	0	0
22″ Flange weld	1120	0	0	0	0
22″ Bolt up	720	0	0	0	0
20″ Exhaust explosion pipe	220	0	0	0	0
20″ Bolt up	80	0	0	0	0
22″ Fixed/sliding support	280	0	0	0	0
22″ Exhaust gas pipe (stack)	560	0	0	0	0
22″ Flange weld	200	0	0	0	0
22″ Bolt up	700	0	0	0	0
20″ Rupture disk	80	0	0	0	0
Support platform at explosion pipe					
Beams	512	0	0	0	0
X-brace	130	0	0	0	0
AB	80	0	0	0	0
Caged ladder	80	0	0	0	0
Platform	80	0	0	0	0
Railing	240	0	0	0	0
Expansion joint	240	0	0	0	0
Equipment installation man-hours	8020	0	0	0	0

7.15   Installation of SCR and urea system (A13) estimate data

Description	MH	Unit
SCR	60	EA
Walking platform (attach to SCR)	40	SF
Silencer	10	EA
Bolt up to silencer	11	EA
26″ Reactor mixing pipe	47	LF
26″ Bolt up	19	EA
Silencers 2354#	10	EA
Silencer support structure with platform		
Beams	2	EA
X-brace	4	EA
AB	3	EA
Caged ladder	10	LF
Platform	10	SF
Railing	4	LF
Column	6	EA
Urea tank 13′ × 36′ 35,700 gal	50	EA
Dress out—filling, suction, drain, vent, overflow	60	EA
Instruments	4	EA
Bolt down	4	EA
Urea pump rack	10	EA
Urea pump control unit	10	EA

7.16   Installation of SCR and urea system (A13) estimate

Description	MH	Qty	Unit	Qty	Unit	BM	IW	PF
SCR and urea system (A13)						0	0	0
SCR	60.0	14	EA	0	EA		0	
Walking platform (attach to SCR)	40.0	14	SF	0	SF		0	
Silencer	10.0	14	EA	0	EA		0	
Bolt up to silencer	11.4	14	EA	0	EA			0
26″ Reactor mixing pipe	47.1	14	LF	0	LF			0
26″ Bolt up	18.6	28	EA	0	EA			0
Silencers 2354#	10.0	14	EA	0	EA		0	
Silencer support structure with platform								
Beams	2.0	64	EA	0	EA		0	
X-brace	4.0	64	EA	0	EA		0	
AB	2.5	16	EA	0	EA		0	
Caged ladder	10.0	4	LF	0	LF		0	
Platform	10.0	12	SF	0	SF		0	
Railing	4.0	36	LF	0	LF		0	
Column	6.3	16	EA	0	EA		0	
Urea tank 13′ × 36′ 35,700 gal	50.0	2	EA	0	EA		0	
Dress out—filling, suction, drain, vent, overflow	60.0	2	EA	0	EA		0	
Instruments	4.0	12	EA	0	EA		0	
Bolt down	4.3	28	EA	0	EA		0	
Urea pump rack	10.0	14	EA	0	EA		0	
Urea pump control unit	10.0	14	EA	0	EA		0	

7.17 SCR and urea system (A13) installation man-hours

Facility-Diesel Power Plant	Actual	Estimate			
Description	MH	BM	IW	PF	MH
SCR	840	0	0	0	0
Walking platform (attach to SCR)	560	0	0	0	0
Silencer	140	0	0	0	0
Bolt up to silencer	160	0	0	0	0
26″ Reactor mixing pipe	660	0	0	0	0
26″ Bolt up	520	0	0	0	0
Silencers 2354#	140	0	0	0	0
Silencer support structure with platform					
Beams	128	0	0	0	0
X-brace	258	0	0	0	0
AB	40	0	0	0	0
Caged ladder	40	0	0	0	0
Platform	120	0	0	0	0
Railing	144	0	0	0	0
Column	100	0	0	0	0
Urea tank 13′ × 36′ 35,700 gal	100	0	0	0	0
Dress out—filling, suction, drain, vent, overflow	120	0	0	0	0
Instruments	48	0	0	0	0
Bolt down	120	0	0	0	0
Urea pump rack	140	0	0	0	0
Urea pump control unit	140	0	0	0	0
Equipment installation man-hours	4518	0	0	0	0

7.18 Installation of cooling system (A10) estimate data

Description	MH	Unit
LT radiator	8.8	EA
VEA 901 maintenance water tank 10890#	20.0	EA
Bolt down	2.0	EA
Dress out—filling, suction, drain, vent, overflow	20.0	EA
VEA 011 expansion vessel 628#	10.0	EA
Instrument air bottle	10.0	EA
Air supply unit	20.0	EA
TSB air supply vessel	20.0	EA
Bolt down	2.0	EA
Support structure		
Legs-columns	2.3	EA
Beams	2.0	EA
Connections	1.5	EA
Bolt down—AB	1.0	EA
Platform	10.0	SF
Handrail at ends	2.0	LF
Handrail at sides	2.0	LF

7.19   Installation of cooling system (A10) estimate

Description	MH	Qty	Unit	Qty	Unit	BM	IW	PF
Cooling system (A10) estimate						0	0	0
LT radiator	8.8	224	EA	0	EA		0	
VEA 901 maintenance water tank 10890#	20.0	1	EA	0	EA		0	
Bolt down	2.0	4	EA	0	EA		0	
Dress out—filling, suction, drain, vent, overflow	20.0	1	EA	0	EA		0	
VEA 011 expansion vessel 628#	10.0	1	EA	0	EA		0	
Instrument air bottle	10.0	14	EA	0	EA		0	
Air supply unit	20.0	2	EA	0	EA		0	
TSB air supply vessel	20.0	4	EA	0	EA		0	
Bolt down	2.0	4	EA	0	EA		0	
Support structure								
Legs-columns	2.3	256	EA	0	EA		0	
Beams	2.0	88	EA	0	EA		0	
Connections	1.5	64	EA	0	EA		0	
Bolt down – AB	1.0	256	EA	0	EA		0	
Platform	10.0	16	SF	0	SF		0	
Handrail at ends	2.0	224	LF	0	LF		0	
Handrail at sides	2.0	56	LF	0	LF		0	

7.20   Cooling system (A10) installation man-hours

Facility—diesel power plant	Actual	Estimate			
Description	MH	BM	IW	PF	MH
LT radiator	1980	0	0	0	0
VEA 901 maintenance water tank 10890#	20	0	0	0	0
Bolt down	8	0	0	0	0
Dress out—filling, suction, drain, vent, overflow	20	0	0	0	0
VEA 011 expansion vessel 628#	10	0	0	0	0
Instrument air bottle	140	0	0	0	0
Air supply unit	40	0	0	0	0
TSB air supply vessel	80	0	0	0	0
Bolt down	8	0	0	0	0
Support structure					
Legs-columns	580	0	0	0	0
Beams	176	0	0	0	0
Connections	96	0	0	0	0
Bolt down—AB	256	0	0	0	0
Platform	160	0	0	0	0
Handrail at ends	448	0	0	0	0
Handrail at sides	112	0	0	0	0
Equipment installation man-hours	4134	0	0	0	0

7.21 Installation of intake air system (A11) estimate data

Description	MH	Unit
Starting air bottle 9555#	40.0	EA
Starting air compressor	40.0	EA
Control and working air compressor unit	40.0	EA
Expansion vessel 1708#	30.0	EA
Charge air filter 2116#	30.0	EA
Control air bottle	40.0	EA
Air receiver 10050#	40.0	EA
Bolt down	2.7	EA
Flex hose	2.0	EA
Bolt up	2.0	EA

7.22 Installation of intake air system (A11) estimate

Description	MH	Qty	Unit	Qty	Unit	BM	IW	PF
Intake air system (A11) estimate						0	0	0
Starting air bottle 9555#	40.0	4	EA	0	EA		0	
Starting air compressor	40.0	2	EA	0	EA		0	
Control and working air compressor unit	40.0	2	EA	0	EA		0	
Expansion vessel 1708#	30.0	14	EA	0	EA		0	
Charge air filter 2116#	30.0	14	EA	0	EA		0	
Control air bottle	40.0	2	EA	0	EA		0	
Air receiver 10050#	40.0	2	EA	0	EA		0	
Bolt down	2.7	12	EA	0	EA		0	
Flex hose	2.0	20	EA	0	EA		0	
Bolt up	2.0	40	EA	0	EA		0	

7.23 Intake air system (A11) installation man hours

Facility—diesel power plant	Actual		Estimate		
Description	MH	BM	IW	PF	MH
Intake air system (A11) estimate					
Starting air bottle 9555#	160	0	0	0	0
Starting air compressor	80	0	0	0	0
Control and working air compressor unit	80	0	0	0	0
Expansion vessel 1708#	420	0	0	0	0
Charge air filter 2116#	420	0	0	0	0
Control air bottle	80	0	0	0	0
Air receiver 10050#	80	0	0	0	0
Bolt down	32	0	0	0	0
Flex hose	40	0	0	0	0
Bolt up	80	0	0	0	0
Equipment installation man-hours	1472	0	0	0	0

7.24 Installation of fire water system (A14) estimate data

Description	MH	Unit
Fire hose equipment	10.0	EA
Hose cabinet	10.0	EA
Fire water hydrant	10.0	EA
Extinguisher	4.0	EA
Mobile foam unit	10.0	EA

7.25 Installation of fire water system (A14) estimate

Description	MH	Qty	Unit	Qty	Unit	BM	IW	PF
Fire water system (A14) estimate						0	0	0
Fire hose equipment	10.0	13	EA	0	EA	0		
Hose cabinet	10.0	3	EA	0	EA	0		
Fire water hydrant	10.0	26	EA	0	EA	0		
Extinguisher	4.0	20	EA	0	EA	0		
Mobile foam unit	10.0	4	EA	0	EA	0		

7.26   Fire water system (A14) installation man hours

Facility—diesel power plant	Actual		Estimate		
Description	MH	BM	IW	PF	MH
Fire hose equipment	130	0	0	0	0
Hose cabinet	30	0	0	0	0
Fire water hydrant	260	0	0	0	0
Extinguisher	80	0	0	0	0
Mobile foam unit	40	0	0	0	0
Equipment installation man-hours	540	0	0	0	0

7.27   Installation of utility gas heating system estimate data

Description	MH	Unit
Gas regulating unit	30.0	FA
Fuel oil unit	20.0	EA
Bolt up	4.0	EA

7.28   Installation of utility gas heating system estimate

Description	MH	Qty	Unit	Qty	Unit	BM	IW	PF
Utility gas heating system estimate						0	0	0
Gas regulating unit	30.0	14	EA	0	EA		0	
Fuel oil unit	20.0	5	EA	0	EA		0	
Bolt up	4.0	10	EA	0	EA		0	

7.29   Utility gas heating system estimate installation man hours

Facility—diesel power plant	Actual		Estimate		
Description	MH	BM	IW	PF	MH
Gas regulating unit	420	0	0	0	0
Fuel oil unit	100	0	0	0	0
Bolt up	40	0	0	0	0
Equipment installation man-hours	560	0	0	0	0

7.30   Installation of lube oil system (pump and tanks) estimate data

Description	MH	Unit
Lube oil service tank 1000 gal	20.0	EA
Dress out—filling, suction, drain, vent, overflow	60.0	EA
Instrument	4.0	EA
Clean lube oil tank 12,000 gal	40.0	EA
Dress out—filling, suction, drain, vent, overflow	60.0	EA
Instrument	4.0	EA
Flushing filter	10.0	EA
Flex rubber bellows	4.0	EA
Bolt up	3.0	EA
Used lube oil tank 4000 gal	20.0	EA
Dress out—filling, suction, drain, vent, overflow	60.0	EA
Instrument	4.0	EA
Pipe bridge platform with stairs	360.0	EA
Pipe supports	10.3	EA
AB (PS)	2.1	EA
Lube oil pump unit	40.0	EA
Bolt down	4.0	EA

7.31   Installation of lube oil system (pump and tanks) estimate

Description	MH	Qty	Unit	Qty	Unit	IW
Lube oil system (pump and tanks)						**0**
Lube oil service tank 1000 gal	20.0	1	EA	0	EA	0
Dress out—filling, suction, drain, vent, overflow	60.0	1	EA	0	EA	0
Instrument	4.0	5	EA	0	EA	0
Clean lube oil tank 12,000 gal	40.0	1	EA	0	EA	0
Dress out—filling, suction, drain, vent, overflow	60.0	1	EA	0	EA	0
Instrument	4.0	6	EA	0	EA	0
Flushing filter	10.0	1	EA	0	EA	0
Flex rubber bellows	4.0	21	EA	0	EA	0
Bolt up	3.0	42	EA	0	EA	0
Used lube oil tank 4000 gal	20.0	1	EA	0	EA	0
Dress out—filling, suction, drain, vent, overflow	60.0	1	EA	0	EA	0
Instrument	4.0	6	EA	0	EA	0
Pipe bridge platform with stairs	360.0	1	EA	0	EA	0
Pipe supports	10.3	33	EA	0	EA	0
AB (PS)	2.1	122	EA	0	EA	0
Lube oil pump unit	40.0	1	EA	0	EA	0
Bolt down	4.0	4	EA	0	EA	0

7.32 Lube oil system (pump and tanks) installation man-hour

Facility—diesel power plant	Actual	Estimate			
Description	MH	BM	IW	PF	MH
Lube oil service tank 1000 gal	20	0	0	0	0
Dress out—filling, suction, drain, vent, overflow	60	0	0	0	0
Instrument	20	0	0	0	0
Clean lube oil tank 12,000 gal	40	0	0	0	0
Dress out—filling, suction, drain, vent, overflow	60	0	0	0	0
Instrument	24	0	0	0	0
Flushing filter	10	0	0	0	0
Flex rubber bellows	84	0	0	0	0
Bolt up	126	0	0	0	0
Used lube oil tank 4000 gal	20	0	0	0	0
Dress out—filling, suction, drain, vent, overflow	60	0	0	0	0
Instrument	24	0	0	0	0
Pipe bridge platform with stairs	360	0	0	0	0
Pipe supports	340	0	0	0	0
AB (PS)	260	0	0	0	0
Lube oil pump unit	40	0	0	0	0
Equipment installation man-hours	1548	0	0	0	0

7.33 Installation of utility water estimate data

Description	MH	Unit
DAD/901 oil water transfer pump 342#	40.0	EA
Bolt down	4.0	EA
NOx controller	4.0	EA
SEN dosing box	4.0	EA
IC pipe	2.0	LF
Bolt up	3.1	EA
Flex rubber bellows	4.0	EA
Bolt up	3.2	EA

7.34   Installation of utility water estimate

Description	MH	Qty	Unit	Qty	Unit	BM	IW	PF
Utility water						0	0	0
DAD/901 oil water transfer pump 342#	40.0	1	EA	0	EA		0	
Bolt down	4.0	4	EA	0	EA		0	
NOx controller	4.0	14	EA	0	EA		0	
SEN dosing box	4.0	14	EA	0	EA		0	
IC pipe	2.0	14	LF	0	LF			0
Bolt up	3.1	9	EA	0	EA			0
Flex rubber bellows	4.0	17	EA	0	EA			0
Bolt up	3.2	34	EA	0	EA			0

7.35   Utility water installation man hour

Facility-diesel power plant	Actual				Estimate
Description	MH	BM	IW	PF	MH
DAD/901 oil water transfer pump 342#	40	0	0	0	0
Bolt down	16	0	0	0	0
NOx controller	56	0	0	0	0
SEN dosing box	56	0	0	0	0
IC pipe	28	0	0	0	0
Bolt up	28	0	0	0	0
Flex rubber bellows	68	0	0	0	0
Bolt up	108	0	0	0	0
Equipment installation man-hours	400	0	0	0	0

7.36   Installation of secondary pipe support estimate data

Description	MH	Unit
Pipe support	8.0	EA
AB	1.0	EA

7.37   Installation of secondary pipe support estimate

Description	MH	Qty	Unit	Qty	Unit	BM	IW	PF
Secondary pipe support						0	0	0
Pipe support	8.0	291	EA	0	EA			0
AB	1.0	1100	EA	0	EA			0

7.38 Secondary pipe support installation man hours

Facility—diesel power plant	Actual		Estimate		
Description	MH	BM	IW	PF	MH
Pipe support	2328	0	0	0	0
AB	1100	0	0	0	0
Equipment installation man-hours	3428	0	0	0	0

7.39 Installation of valves list estimate data

Description	MH	Unit
8″ In-line valve flange	11.6	EA
0.5″ In-line valve NPT	3.8	EA
3″ In-line valve flange	4.4	EA
1″ In-line valve NPT	1.6	EA
2″ In-line valve sw	2.6	EA
1″ In-line valve sw	2.6	EA
2.5″ In-line valve flange	3.6	EA
1.5″ In-line valve sw	2.6	EA
0.375″ In-line valve NPT	1.8	EA
0.75″ In-line valve sw	2.6	EA
0.75″ In-line valve NPT	1.8	EA
4″ In-line valve flange	5.8	EA
6″ In-line valve flange	8.7	EA
1.5″ In-line valve bw	2.6	EA
2″ In-line valve bw	2.6	EA
2″ In-line valve flg'd	2.2	EA

7.40   Installation of valves list estimate

Description	MH	Qty	Unit	Qty	Unit	BM	IW	PF
Valves list						0	0	0
8″ In-line valve flange	11.6	2	EA	0	EA			0
0.5″ In-line valve NPT	3.8	61	EA	0	EA			0
3″ In-line valve flange	4.4	46	EA	0	EA			0
1″ In-line valve NPT	1.6	29	EA	0	EA			0
2″ In-line valve sw	2.6	16	EA	0	EA			0
1″ In-line valve sw	2.6	10	EA	0	EA			0
2.5″ In-line valve flange	3.6	5	EA	0	EA			0
1.5″ In-line valve sw	2.6	2	EA	0	EA			0
0.375″ In-line valve NPT	1.8	14	EA	0	EA			0
0.75″ In-line valve sw	2.6	14	EA	0	EA			0
0.75″ In-line valve NPT	1.8	1	EA	0	EA			0
4″ In-line valve flange	5.8	56	EA	0	EA			0
6″ In-line valve flange	8.7	9	EA	0	EA			0
1.5″ In-line valve bw	2.6	4	EA	0	EA			0
2″ In-line valve bw	2.6	4	EA	0	EA			0
2″ In-line valve flg'd	2.2	4	EA	0	EA			0

7.41   Valves list installation man-hour

Facility—diesel power plant	Actual		Estimate		
Description	MH	BM	IW	PF	MH
8″ In-line valve flanged	23	0	0	0	0
5″ In-line valve NPT	232	0	0	0	0
3″ In-line valve flanged	200	0	0	0	0
1″ In-line valve NPT	46	0	0	0	0
2″ In-line valve sw	42	0	0	0	0
1″ In-line valve sw	26	0	0	0	0
2.5″ In-line valve flanged	18	0	0	0	0
1.5″ In-line valve sw	5	0	0	0	0
0.375″ In-line valve NPT	25	0	0	0	0
0.75″ In-line valve sw	36	0	0	0	0
0.75″ In-line valve NPT	2	0	0	0	0
4″ In-line valve flanged	325	0	0	0	0
6″ In-line valve flanged	78	0	0	0	0
1.5″ In-line valve bw	10	0	0	0	0
2″ In-line valve bw	10	0	0	0	0
2″ In-line valve flanged	9	0	0	0	0
Equipment installation man-hours	1089	0	0	0	0

7.42 Installation of manometer/thermometer list estimate data

Description	MH	Unit
Manometer	2.0	EA
Thermometer	2.0	EA

7.43 Installation of manometer/thermometer list estimate

Description	MH	Qty	Unit	Qty	Unit	BM	IW	PF
Manometer/thermometer list						0	0	0
Manometer	2.0	60	EA	0	EA			0
Thermometer	2.0	56	EA	0	EA			0

7.44 Manometer/thermometer list installation man hour

Facility—diesel power plant	Actual		Estimate		
Description	MH	BM	IW	PF	MH
Manometer	120	0	0	0	0
Thermometer	112	0	0	0	0
Equipment installation man-hours	232	0	0	0	0

7.45 Leaking channel set estimate data

Description	MH	Unit
Leaking channel WDAAA373661 1 set b	60.0	EA
Leaking channel 2 WDAAA373663 2 set b	60.0	EA
Leaking channel WDAAA373662 1 set b	80.0	EA

7.46 Leaking channel set estimate

Description	MH	Qty	Unit	Qty	Unit	BM	IW	PF
Leaking channel set						0	0	0
Leaking channel WDAAA373661 1 set b	60.0	1	EA	0	EA		0	
Leaking channel 2 WDAAA373663 2 set b	60.0	2	EA	0	EA		0	
Leaking channel WDAAA373662 1 set b	80.0	1	EA	0	EA		0	

7.47 Leaking channel set installation man hour

Facility—diesel power plant	Actual		Estimate		
Description	MH	BM	IW	PF	MH
Leaking channel WDAAA373661 1 set b	60	0	0	0	0
Leaking channel 2 WDAAA373663 2 set b	120	0	0	0	0
Leaking channel WDAAA373662 1 set b	80	0	0	0	0
Equipment installation man-hours	260	0	0	0	0

7.48 Man hour breakdown

7.48.1 Diesel power plant

Owner-furnished equipment

Direct craft man-hours	Actual	BM	IW	MW	Estimate
Scope of work	MH	MH	MH	MH	MH
4.6 Engines, couplings, and generators (B2) installation man-hours	4640	0	0	0	0
4.9 Engines and aux platforms (A3) installation man-hours	4918	0	0	0	0
4.12 Installation of pipe modules (B3) installation man-hours	1920	0	0	0	0
4.15 Exhaust gas system (A11) installation man-hours	8020	0	0	0	0
4.18 SCR and urea system (A13) installation man-hours	4518	0	0	0	0
4.21 Cooling system (A10) installation man-hours	4134	0	0	0	0
4.24 Intake air system (A11) installation man-hours	1472	0	0	0	0
4.27 Fire water system (A14) installation man-hours	540	0	0	0	0
4.30 Utility gas heating system estimate installation man-hours	560	0	0	0	0
4.33 Lube oil system (pump and tanks) installation man-hour	1548	0	0	0	0
4.36 Utility water installation man-hour	400	0	0	0	0
4.39 Secondary pipe support installation man-hours	3428	0	0	0	0
4.42 Valves list installation man-hour	1089	0	0	0	0
4.45 Manometer/thermometer list installation man-hour	232	0	0	0	0
4.48 Leaking channel set installation man-hour	260	0	0	0	0

Chapter 8

Coal-fired power plant

8.1 Equipment descriptions

8.1.1 Equipment descriptions—Coal-fired boiler pressure parts sheet 1

Coal-fired power plants produce electricity by burning coal in a boiler to produce steam. The steam produced, under pressure, flows into a turbine, which spins a generator to create electricity. The steam is then cooled, condensed back into water, and returned to the boiler to start the process over.

The project includes the installation of a new coal-fired boiler and the installation of new auxiliary equipment. Erection of the boiler, selective catalytic reduction (SCR) system, air heater, FD and primary air fans, flue work, fly ash handling system, electrostatic precipitator, ID fan, coal silo and pulverize, bottom ash handling system, ductwork and breeching, structural steel, and boiler casing

Coal-Fired Boiler
Steam pressure:
1800–2600 psi
Steam temperature:
1000–1050°F (538–566°C)
Fuel:
Pulverized coal

Industrial Process Plant Construction Estimating and Man-Hour Analysis.
https://doi.org/10.1016/B978-0-12-818648-0.00008-9

8.1.2 Equipment descriptions—Coal-fired boiler pressure parts sheet 2

Boiler-Pressure Parts
Attemperators (first and second stage)
Steam drum with internals
Economizer sections and headers
Membrane furnace wall panels and individual tubes
Primary and secondary superheater sections and headers
Reheat superheater sections and headers
Furnace arch tubes
Extended surface tube sections
Tube wall panels, partial panels, tube panel inserts
Main steam line
Down comers, supply, and riser tubes
Platen secondary superheaters
Intermediate and final secondary superheaters

8.1.3   Equipment descriptions—Auxiliary equipment

Coal Silo (KD)
Coal Pulverize
Pulverize, gear drive unit, and housing shipped separately (field installation)
Belt and coupling guards
Base plates
Feeders, separately mounted
Feeder controller
Primary air fan
Air seal blower and piping
Pyrites discharge
Pressure lube system
Air Heater
Tube sheets
Baffle sheets—turn and vibration
Supports—tube sheets
Erect grid support steel
Supports—main
Install tubes
Expand 2″ ends
Expand 2-1/2″ ends
Casing, veneer including batten bars and channels
Casing, main and hoppers including expansion joints and doors
Dampers
Selective Catalytic Reduction (SCR)
SCR reactor
Assemble shell
Field joint
Ammonia grid module
Field joint
Remove shipping angles
Install tie plate
Seal frame
Catalyst modules
Install catalyst

8.1.4   Equipment descriptions—Fan system sheet 1

FANS

Perform all millwright work for the four fans. Alignment and grouting
Install sole and base plates
Attach fan to foundation—bolted connection

FD Fan

Set fan housing split
Set impeller assembly/shaft
Alignment fan bearings
Seal weld housing inside/outside
Weld inlet cones and stud rings
Weld shaft seal stud rings
Coupling alignment
Install damper

ID Fan

Set fan housing split
Set impeller assembly/shaft
Alignment fan bearings
Seal weld housing inside/outside
Weld inlet cones and stud rings
Weld shaft seal stud rings
Coupling alignment
Install damper

8.1.5   Equipment descriptions—Fan system sheet 2

Primary air fan
Set fan housing split
Set impeller assembly/shaft
Alignment fan bearings
Seal weld housing inside/outside
Weld inlet cones and stud rings
Weld shaft seal stud rings
Coupling alignment
Install damper
Pulverized seal air fan
Set fan housing split
Set impeller assembly/shaft
Alignment fan bearings
Seal weld housing inside/outside
Weld inlet cones and stud rings
Weld shaft seal stud rings
Coupling alignment
Install damper

8.1.6 Equipment descriptions—Electrostatic precipitator sheet 1

Casing

Prefabricate wall and partition panels

Install anvil beams and antisneak age baffles on end frames/roof beams and bottom baffles

Set partition and wall panels; stabilize with top-end frames and bottom-end frames and intermittently with intermediate roof beams and supporting bottom baffles

Install remaining intermediate roof beams and supporting bottom baffles

Square and plumb casing and began weld out

Complete casing welding, including the hoppers

Drop in preassembled hoppers with lower alignment frames and scaffolding inside

Secure interior cross ties between roof beams, end frames, and bottom baffles

Hang inlet and outlet perforated plates

Preassemble and mount inlet and outlet plenums

Install inlet and outlet plenum cone sections

Install platforms, stairways, and ladders

Internals

Lift and set collecting plates into ESP

Join plate halves and attach plates to anvil beams

Do preliminary alignment of collecting plates

Preassemble discharge electrodes to high voltage frames

Lift high voltage frames and rest on collecting plates

8.1.7 Equipment descriptions—Electrostatic precipitator sheet 2

Penthouse
Preassemble hot roof sections
Install hot roof sections
Install lower alignment frames to discharge electrodes
Suspend high voltage frames from support insulators
Complete welding of penthouse walls and hot roof
Weld out rapper sleeves (installed below)
ESP Roof
Preassemble cold roof sections
Insulate underside of cold roof
Set cold roof sections and expansion joints
Install roof handrail
Install and weld out rapper sleeves
Weld out cold roof
Install rapper shafts, rapper adjusting studs, and rappers
Align rappers
Set TR sets
Install duct support steel
Install inlet ductwork on the support steel including the louver dampers
Install seal air system platform on duct support steel
Set seal air fan and heater
Support structure
Sliding plates and cap plates

8.1.8 Equipment descriptions—Structural steel and boiler casing

Structural Steel
Erect Boiler Structural Steel (Stairs, Platforms, Grating & Handrails)
Main steel
Platform framing
Grating
Handrail
Erect Penthouse Steel
Install Platform, Grating & Handrail
Platform framing
Grating
Handrail
Install Stair Nos. 1 and 2
Structural steel
Platform framing
Grating
Handrail
Stair treads
Erect Casing -Generating Bank, Boiler and Penthouse

8.1.9 Equipment descriptions—Ductwork and breeching and bottom ash handling system

Ductwork and Breeching

Install ductwork and breeching and expansion joints. Fit up and seal welding from inside. Fit up flanges tack welded to sections to install and bolt field joints.

Ductwork will rest on sliding supports or hang from supports.

Bottom Ash Handling System

Equipment consists of refractory-lined bottom ash chute and one (1) submerged ash drag chain conveyor.

Bottom ash chute

Submerged ash drag conveyor with outlet chute.

Shipped in sections:

Head section—motor and drive shipped attached to head

Neck section

Transition section

Straight section

Tail section—take up shipped assembled

Scope of work:

Assembly of housing

Assembly of drag chain

Installation of drive units

Pneumatic/hydraulic take-up units, including air piping

Discharge chutes

Grouting

Refractory-lined chute on inlet of submerged ash drag

All water piping to and from submerged ash drag and drains

Coal-Fired Boiler

8.2 General scope of field work required for each coal-fired boiler sheet 1

Scope of Work-Field Erection
Pressure Parts
Install Drum, Erect Headers & Panels, Furnace Wall & Roof Tubes, Weld Tubes & Panels
Install Steam Drum
Drums include straps and/or U-bolts: steam drum (200,001 lb ≤ Weight ≤ 300,000 lb)
Steam drum trim piping
Drum internals—remove and reinstall
Set rigging for hoisting assemblies
Hoist into place
Headers, Furnace and Water wall Panels
Boiler Sidewall Tubes
Headers—loose
Wall panels—shop assembled without headers
Weld tubes—panel to panel—(2″< BW ≤ 2-1/2″ TIG) (2001–2600) psi
Fit and weld filler bar at tube welds
Weld tubes to header stubs—(2″< BW ≤ 2-1/2″ TIG) (2001–2600) psi
Fit and weld filler joining membrane panels
Furnace Front Wall Tubes
Headers—loose
Wall panels—shop assembled without headers
Weld tubes—panel to panel—(2″ < BW ≤ 2-1/2″ TIG) (2001–2600) psi
Fit and weld filler bar at tube welds
Weld Tubes to header stubs—(2″ < BW ≤ 2-1/2″ TIG) (2001–2600) psi
Fit and weld filler joining membrane panels

8.3 General Scope of Field Work Required for Each Coal-Fired Boiler Sheet 2

Furnace Roof/Floor Tubes

Headers—loose

Tubes—loose, furnace including screen

Weld tubes to header stubs—(2-1/2″ < BW ≤ 3″ TIG) (2001—2600) psi

Furnace Sidewall Tubes

Headers—loose

Wall Panels—shop assembled without headers

Weld Tubes—panel to panel—(2″< BW ≤ 2-1/2″ TIG) (2001–2600) psi

Fit and weld filler bar at tube welds

Weld tubes to header stubs—(2″< BW ≤ 2-1/2″ TIG) (2001–2600) psi

Fit and weld filler joining membrane panels

Furnace Rear Wall Tubes

Furnace rear wall tubes (wall panel—membrane)

Furnace rear wall tubes (loose)

Weld tubes—panel to panel—(2″< BW ≤ 2-1/2″ TIG) (2001–2600) psi

Fit and weld filler bar at tube welds

Weld tubes to header stubs—(2″< BW ≤ 2-1/2″ TIG) (2001–2600) psi

Fit and weld filler joining membrane panels

Install and weld membrane and scallop bars on nose tubes

8.4 General scope of field work required for each coal-fired boiler sheet 3

Scope of Work-Field Erection
Pressure Parts
Install and Weld Attemperators, Superheater, and Reheat Components
Final Secondary Superheater
Tubes—connecting
Tubes—sections
Headers
Weld tubes to header stubs—(3-1/2″< BW ≤ 4″ TIG) (2001–2600) psi
Fit and weld filler bar at tube welds
First and Second Stage Attemperators
Spray type—complete with body and internals (excludes welding)
Drum type—complete with internals, piping, valves, and supports (excludes welding)
Connecting tubes (4-1/2″ welded to stub above and 4″ expand in drum, weld to header)
4″ OD, expand tubes in drums (2000 psi)
Weld Tubes to header stubs—(4″< BW ≤ 4-1/2″ TIG) (2001–2600) psi

Intermediate Secondary Superheater-Pendant/Inverted Loop
Headers
Supports—rods
Tube sections
Tube section supports
Saturated connections
Mountings Seals
Seals
Weld tubes to header stubs—(3-1/2″< BW ≤ 4″ TIG) (2001–2600) psi

8.5 General scope of field work required for each coal-fired boiler sheet 4

Platen Secondary Superheater-Pendant/Inverted Loop
Headers
Supports—rods
Tube sections
Tube section supports
Saturated connections
Mountings seals
Seals
Weld tubes to header stubs—(3-1/2″< BW ≤ 4″ TIG) (2001–2600) psi
Reheat Superheater-Pendant/Inverted Loop
Headers
Supports—rods
Tube sections
Tube section supports
Saturated connections
Mounting seals
Seals
Weld tubes to header stubs—(3-1/2″< BW ≤ 4″ TIG) (2001–2600) psi
Primary Superheater-Pendant/Inverted Loop
Headers
Supports—rods
Tube sections
Tube section supports
Saturated connections
Mounting seals
Seals
Weld tubes to header stubs—(3-1/2″< BW ≤ 4″ TIG) (2001–2600) psi

8.6   General scope of field work required for each coal-fired boiler sheet 5

Scope of Work-Field Erection
Pressure Parts
Supply and Riser Tubes, Down comers, Boiler Equipment, Burners, Economizer and Diamond Soot blowers
Bare Tube Economizer
Tube sections
Headers
Integral support steel
Set rigging for hoisting assemblies
Hoist into place
Weld tubes to header stubs—($2'' <$ Ring Weld $\leq$ 2-1/2$''$) SMAW (2001–2600) psi
Fit and weld filler bar at tube welds
Casing and hoppers including doors
Supply and Riser Tubes
Headers
Tubes—make up or supply
Tubes—riser
Weld tubes to header stubs—($4'' <$ BW $\leq$ 4-1/2$''$ TIG) (2001–2600) psi
Down Comers
Install down comer
Diameter weld, girth
Diameter weld, ring
Diameter weld, ring

8.7 General scope of field work required for each coal-fired boiler sheet 6

Boiler Equipment

Floor and hopper support steel
Suspension steel
Backstays
Erect casing panels and assemblies
Fit and weld casing seams
Burner wine box—primary, secondary, and tertiary
Hoppers
Doors and frames
Seals
Expanded metal lathe
Closure plates
Coal, volumetric, or gravimetric feeders

Burners

DRB-4Z low No_χ coal burner
Dual zone No_χ port (overfired air system)
Inter tube burners include blocks, tips, and riffle casings
Automatic lighters include welding

Diamond Boiler Cleaning Equipment—Retractable Soot blower

Soot blower SB, SB w/puff
Pressure reducing station
Trays and channels for supporting tubing
Tubing for automatic sequential air operation
Air compressor and/or receiver
Auto sequential panel or air master controller
IK structural supports

8.8   Coal-fired boiler estimate data

8.8.1   Install drum, erect headers, panels, furnace wall, roof tubes, weld tubes/panels sheet 1

Description	MH	Unit
Install steam drum		
Drums—include straps and/or U-bolts: steam drum (200,001 lb ≤ Weight ≤ 300,000 lb)	9.00	Ton
Steam drum trim piping	160.00	Boiler
Drum internals—remove and reinstall	136.00	Boiler
Set rigging for hoisting assemblies	60.00	Boiler
Hoist into place	1.50	Ton
Erect headers and wall panels; fit and weld water wall panels		
Boiler sidewall, furnace front wall, roof, sidewall, and rear wall tubes		
Headers—loose	15.00	Ton
Wall panels—shop assembled without headers	10.00	Ton
Weld tubes—panel to panel—(2″ < BW ≤ 2-1/2″ TIG) (2001–2600) psi	4.20	EA
Fit and weld filler bar at tube welds	1.00	Space
Weld tubes to header stubs—(2″< BW ≤ 2-1/2″ TIG) (2001–2600) psi	4.20	EA
Fit and weld filler joining membrane panels	0.70	LF
Tubes—loose, furnace including screen	1.40	EA
Weld tubes to header stubs—(2-1/2″ < BW ≤ 3″ TIG) (2001–2600) psi	4.60	EA
Furnace rear wall tubes (wall panel—membrane)	12.00	Ton
Install and weld membrane and scallop bars on nose tubes	2.30	Ton

8.8.2   Install and weld attemperators, superheater, and reheat components sheet 2

Description	MH	Unit
Install and weld attemperators, superheater, and reheat components		
Tubes—connecting	1.65	EA
Tubes—sections—final secondary superheater	7.00	Ton
Headers	15.00	Ton
Weld tubes to header stubs—(3-1/2″< BW ≤ 4″ TIG) (2001–2600) psi	5.90	EA
Fit and weld filler bar at tube welds	1.00	Space
Spray type—complete with body and internals (excludes welding)	20.00	Ton
Drum type—complete with internals, piping, valves, and supports (excludes welding)	32.00	Ton
4″ OD, expand tubes in drums (1800 psi)	4.00	EA
Weld tubes to header stubs—(4″< BW ≤ 4-1/2 TIG) (2001–2500) psi	6.50	EA
Headers	65.00	Ton
Supports—rods	65.00	Ton
Tube sections—platen secondary superheater-pendant/inverted loop	20.00	Ton
Tube section supports	20.00	Ton
Saturated connections	35.00	Ton
Mounting seals	90.00	Ton
Seals	175.00	Ton

8.8.3 Tubes, down comers, accessories, burner, soot blowers, economizer sheet 3

Description	MH	Unit
Headers	15.00	Ton
Tubes—make up or supply	2.00	EA
Tubes—riser	1.65	EA
Weld tubes to header stubs—(4″ < BW ≤ 4-1/2 TIG) (2001–2600) psi	6.50	EA
Install down comer	24.00	Ton
Diameter weld, girth—field pipe weld	1.20	DI
Diameter weld, girth—field pipe weld	1.20	DI
Diameter weld, girth—field pipe weld	1.20	DI
Floor and hopper support steel	28.00	Ton
Suspension steel	30.00	Ton
Backstays	36.00	Ton
Erect casing panels and assemblies	15.00	EA
Fit and weld casing seams	0.55	LF
Burner wine box—primary, secondary, and tertiary	75.00	Ton
Hoppers	95.00	Ton
Doors and frames	10.00	EA
Seals	190.00	Ton
Expanded metal lathe	0.08	SF
Closure plates	95.00	Ton
Coal, volumetric, or gravimetric feeders	16.00	Ton
DRB-4Z low No$_x$ coal burner	240.00	EA
Dual zone No$_x$ port (overfired air system)	200.00	EA
Inter tube burners include blocks, tips, and riffle casings	18.00	Ton

8.8.4 Tubes, down comers, accessories, burner, soot blowers, economizer sheet 4

Description	MH	Unit
Automatic lighters include welding	6.00	EA
Soot blower SB, SB with PVF	151.00	EA
Pressure reducing station	70.00	EA
Trays and channels for supporting tubing	18.00	EA
Tubing for automatic sequential air operation	40.00	EA
Air compressor and/or receiver	18.00	Ton
Auto sequential panel or air master controller	86.00	Panel
IK structural supports	6.00	Blower
Tube sections	7.00	Ton
Headers	15.00	Ton
Integral support steel	28.00	Ton
Set rigging for hoisting assemblies	60.00	Boiler
Hoist into place	1.50	Ton
Weld tubes to header stubs—(2″< BW ≤ 2-1/2″ SMAW (2001–2600) psi	3.10	EA
Fit and weld filler bar at tube welds	1.00	Space
Casing and hoppers including doors	55.00	Ton

8.9 Coal-fired boiler installation estimate

8.9.1 Drum, headers, panels, furnace wall, roof tubes, weld tubes/ panels sheet 1

| | Installation man-hour | | | | |
| | Historical | | | Estimate | |
	MH	Unit	Qty	Unit	BM
Install steam drum					0
Steam drum (200,001 lb ≤ Weight ≤ 300,000 lb)	9.00	Ton	0	Ton	0
Steam drum trim piping	160.00	Boiler	0	Boiler	0
Drum internals—remove and reinstall	136.00	Boiler	0	Boiler	0
Set rigging for hoisting assemblies	60.00	Boiler	0	Boiler	0
Hoist into place	1.50	Ton	0	Ton	0
Headers, furnace, and water wall panels					**0**
Boiler sidewall tubes					
Headers—loose	15.00	Ton	0	Ton	0
Wall Panels—shop assembled without headers	10.00	Ton	0	Ton	0
Weld Tubes—panel to panel—(2″ < BW ≤ 2-1/2″ TIG) (2001–2600) psi	4.20	EA	0	EA	0
Fit and weld filler bar at tube welds	1.00	Space	0	Space	0
Weld tubes to header stubs—(2″ < BW ≤ 2-1/2″ TIG) (2001–2600) psi	4.20	EA	0	EA	0
Fit and weld filler joining membrane panels	0.70	LF	0	LF	0
Furnace front wall tubes					
Headers—loose	15.00	Ton	0	Ton	0
Wall Panels—shop assembled without headers	10.00	Ton	0	Ton	0
Weld tubes—panel to panel—(2″ < BW ≤ 2-1/2″ TIG) (2001–2600) psi	4.20	EA	0	EA	0
Fit and weld filler bar at tube welds	1.00	Space	0	Space	0
Weld tubes to header stubs—(2″ < BW ≤ 2-1/2″ TIG) (2001–2600) psi	4.20	EA	0	EA	0
Fit and weld filler joining membrane panels	0.70	LF	0	LF	0

8.9.2 Drum, headers, panels, furnace wall, roof tubes, weld tubes/ panels sheet 2

	Installation man-hour				
	Historical			Estimate	
	MH	Unit	Qty	Unit	BM
Furnace roof/floor tubes					
Headers—loose	15.00	Ton	0	Ton	0
Tubes—loose, furnace including screen	1.40	EA	0	EA	0
Weld tubes to header stubs—(2-1/2″ < BW ≤ 3″ TIG) (2001–2600) psi	4.60	EA	0	EA	0
Furnace sidewall tubes					
Headers—loose	15.00	Ton	0	Ton	0
Wall Panels—shop assembled without headers	10.00	Ton	0	Ton	0
Weld tubes—panel to panel—(2″ < BW ≤ 2-1/2″ TIG) (2001–2600) psi	4.20	EA	0	EA	0
Fit and weld filler bar at tube welds	1.00	Space	0	Space	0
Weld tubes to header stubs—(2″< BW ≤ 2-1/2″ TIG) (2001–2600) psi	4.20	EA	0	EA	0
Fit and weld filler joining membrane panels	0.70	LF	0	LF	0
Furnace rear wall tubes					
Furnace rear wall tubes (wall panel—membrane)	12.00	Ton	0	Ton	0
Furnace rear wall tubes (loose)	1.40	EA	0	EA	0
Weld tubes—panel to panel—(2″ < BW ≤ 2-1/2″ TIG) (2001–2600) psi	4.20	EA	0	EA	0
Fit and weld filler bar at tube welds	1.00	Space	0	Space	0
Weld tubes to header stubs—(2″ < BW ≤ 2-1/2″ TIG) (2001–2600) psi	4.20	EA	0	EA	0
Fit and weld filler joining membrane panels	0.70	LF	0	LF	0
Install and weld membrane and scallop bars on nose tubes	2.30	Ton	0	Ton	0

8.9.3 Install and weld attemperators, superheater, and reheat components sheet 3

	Installation man-hour				
	Historical			Estimate	
	MH	Unit	Qty	Unit	BM
Install and weld attemperators, superheater, and reheat components					0
Final secondary superheater					
Tubes—connecting	1.65	EA	0	EA	0
Tubes—sections	7.00	Ton	0	Ton	0
Headers	15.00	Ton	0	Ton	0
Weld tubes to header stubs—(3-1/2″ < BW ≤ 4″ TIG) (2001–2600) psi	5.90	EA	0	EA	0
Fit and weld filler bar at tube welds	1.00	Space	0	Space	0
First and second stage attemperators					
Spray type—complete with body and internals (excludes welding)	20.00	Ton	0	Ton	0
Drum type—complete with internals, p&v, support (excludes welding)	32.00	Ton	0	Ton	0
Tubes (4-1/2″ welded to stub and 4″ expand in drum, weld to hdr)					
4″ OD, expand tubes in drums (1800 psi)	4.00	EA	0	EA	0
Weld tubes to header stubs—(4″ < BW ≤ 4-1/2 TIG) (2001–2500) psi	6.50	EA	0	EA	0
Platen secondary superheater-pendant/inverted loop					
Headers	65.00	Ton	0	Ton	0
Supports—rods	65.00	Ton	0	Ton	0
Tube sections	20.00	Ton	0	Ton	0
Tube section supports	20.00	Ton	0	Ton	0
Saturated connections	35.00	Ton	0	Ton	0
Mounting seals	90.00	Ton	0	Ton	0
Seals	175.00	Ton	0	Ton	0
Weld tubes to header stubs—3-1/2″ < BW ≤ 4″ TIG) (2001–2600) PSI	5.90	EA	0	EA	0

8.9.4 Install and weld attemperators, superheater, and reheat components sheet 4

	Installation man-hour				
	Historical			Estimate	
	MH	Unit	Qty	Unit	BM
Intermediate secondary superheater-pendant/inverted loop					
Headers	65.00	Ton	0	Ton	0
Supports—rods	65.00	Ton	0	Ton	0
Tube sections	20.00	Ton	0	Ton	0
Tube section supports	20.00	Ton	0	Ton	0
Saturated connections	35.00	Ton	0	Ton	0
Mounting seals	90.00	Ton	0	Ton	0
Seals	175.00	Ton	0	Ton	0
Weld tubes to header stubs—(3-1/2″ < BW ≤ 4″ TIG) (2001–2600) psi	5.90	EA	0	EA	0
Reheat superheater-pendant/inverted loop					
Headers	65.00	Ton	0	Ton	0
Supports—rods	65.00	Ton	0	Ton	0
Tube sections	20.00	Ton	0	Ton	0
Tube section supports	20.00	Ton	0	Ton	0
Saturated connections	35.00	Ton	0	Ton	0
Mounting seals	90.00	Ton	0	Ton	0
Seals	175.00	Ton	0	Ton	0
Weld tubes to header stubs—(3-1/2″ < BW ≤ 4″ TIG) (2001–2600) psi	5.90	EA	0	EA	0
Primary superheater-pendant/inverted loop					
Headers	65.00	Ton	0	Ton	0
Supports—rods	65.00	Ton	0	Ton	0
Tube sections	20.00	Ton	0	Ton	0
Tube section supports	20.00	Ton	0	Ton	0
Saturated connections	35.00	Ton	0	Ton	0
Mounting seals	90.00	Ton	0	Ton	0
Seals	175.00	Ton	0	Ton	0
Weld tubes to header stubs—(3-1/2″ < BW ≤ 4″ TIG) (2001–2600) psi	5.90	EA	0	EA	0

8.9.5 Tubes, down comers, accessories, burner, soot blowers, economizer sheet 5

	Installation man-hour				
	Historical			Estimate	
	MH	Unit	Qty	Unit	BM
Tubes, down comers, boiler equipment, burners, and diamond soot blowers					0
Supply and riser tubes					
Headers	15.00	Ton	0	Ton	0
Tubes—make up or supply	2.00	EA	0	EA	0
Tubes—Riser	1.65	EA	0	EA	0
Weld tubes to header stubs—(4″ < BW ≤ 4-1/2 TIG) (2001–2600) psi	6.50	EA	0	EA	0
Down comers					
Install down comer	24.00	Ton	0	Ton	0
Diameter weld, girth—field pipe weld	1.20	DI	0	DI	0
Diameter weld, girth—field pipe weld	1.20	DI	0	DI	0
Diameter weld, girth—field pipe weld	1.20	DI	0	DI	0
Boiler equipment					
Floor and hopper support steel	28.00	Ton	0	Ton	0
Suspension steel	30.00	Ton	0	Ton	0
Backstays	36.00	Ton	0	Ton	0
Erect casing panels and assemblies	15.00	EA	0	EA	0
Fit and weld casing seams	0.55	LF	0	LF	0
Burner wine box—primary, secondary, and tertiary	75.00	Ton	0	Ton	0
Hoppers	95.00	Ton	0	Ton	0
Doors and frames	10.00	EA	0	EA	0
Seals	190.00	Ton	0	Ton	0
Expanded metal lathe	0.08	SF	0	SF	0
Closure plates	95.00	Ton	0	Ton	0
Coal, volumetric, or gravimetric feeders	16.00	Ton	0	Ton	0

8.9.6   Tubes, down comers, accessories, burner, soot blowers, economizer sheet 6

	Installation man-hour				
	Historical			Estimate	
	MH	Unit	Qty	Unit	BM
Burners					
DRB-4Z low No$_x$ coal burner	240.00	EA	0	EA	0
Dual zone No$_x$ port (over fire air system)	200.00	EA	0	EA	0
Inter tube burners include blocks, tips, and riffle casings	18.00	Ton	0	Ton	0
Automatic lighters include welding	6.00	EA	0	EA	0
Diamond boiler cleaning equipment—retractable soot blower					
Soot blower SB, SB with PVF	151.00	EA	0	EA	0
Pressure reducing station	70.00	EA	0	EA	0
Trays and channels for supporting tubing	18.00	EA	0	EA	0
Tubing for automatic sequential air operation	40.00	EA	0	EA	0
Air compressor and/or receiver	18.00	Ton	0	Ton	0
Auto sequential panel or air master controller	86.00	Panel	0	Panel	0
IK structural supports	6.00	Blower	0	Blower	0
Bare tube economizer					
Tube sections	7.00	Ton	0	Ton	0
Headers	15.00	Ton	0	Ton	0
Integral support steel	28.00	Ton	0	Ton	0
Set rigging for hoisting assemblies	60.00	Boiler	0	Boiler	0
Hoist into place	1.50	Ton	0	Ton	0
Weld tubes to her stubs—($2'' <$ BW $\leq$ 2-1/2$''$ SMAW)	3.10	EA	0	EA	0
Fit and weld filler bar at tube welds	1.00	Space	0	Space	0
Casing and hoppers including doors	55.00	Ton	0	Ton	0

8.10   Coal-fired boiler installation man-hours

	Estimated	Actual
	BM	BM
Install steam drum	0	0
Headers, furnace, and water wall panels	0	0
Install and weld attemperators, superheater, and reheat components	0	0
Tubes, down comers, boiler equipment, burners and diamond soot blowers	0	0
Coal-fired boiler installation man-hours	0	0

8.11 General scope of field work required for each coal silo and coal pulverizer coal silo

Coal Pulverizer

Pulverizer, gear drive unit, and housing shipped separately (field installation)
Belt and coupling guards
Base plates
Feeders, separately mounted
Feeder controller
Primary air fan
Air seal blower and piping
Pyrites discharge
Pressure lube system

8.12 Coal silo and pulverizer estimate data

8.12.1 Install coal silo and pulverizer sheet 1

Description	MH	Unit
Pulverizer, gear drive unit, and housing (field installation)	8.00	Ton
Belt and coupling guards	9.00	EA
Base plates	30.00	Ton
Feeders, separately mounted	45.00	EA
Feeder controller	28.00	Ton
Primary air fan	48.00	Ton
Air seal blower and piping	72.00	Pulverizer
Pyrites discharge	30.00	Ton
Pressure lube system	86.00	Pulverizer
Coal silo/bunker (KD)	20.00	Ton

8.13 Coal silo and equipment installation estimate

8.13.1 Install coal silo and pulverizer sheet 1

| | Installation man-hour | | | | |
| | Historical | | | Estimate | |
	MH	Unit	Qty	Unit	BM
Install coal silo and coal pulverizer					**0**
Pulverizer, gear drive unit, and housing (field installation)	8.00	Ton	0	Ton	0
Belt and coupling guards	9.00	EA	0	EA	0
Base plates	30.00	Ton	0	Ton	0
Feeders, separately mounted	45.00	EA	0	EA	0
Feeder controller	28.00	Ton	0	Ton	0
Primary air fan	48.00	Ton	0	Ton	0
Air seal blower and piping	72.00	Pulverizer	0	Pulverizer	0
Pyrites discharge	30.00	Ton	0	Ton	0
Pressure lube system	86.00	Pulverizer	0	Pulverizer	0
Coal silo (KD)	20.00	Ton	0	Ton	**0**

8.14 Coal silo and pulverizer installation man hours

| | Estimated | Actual |
	BM	BM
Install coal silo and coal pulverizer	0	0
Coal silo (KD)	0	0
Coal silo and pulverizer installation man-hours	0	0

8.15 General scope of field work required for each air heater

Tube sheets
Baffle sheets—turn and vibration
Supports—tube sheets
Erect grid support steel
Supports—main
Install tubes
Expand 2″ ends
Expand 2-1/2″ ends
Casing, veneer including batten bars and channels
Casing, main and hoppers including expansion joints and doors
Dampers

8.16   Air heater estimate data

8.16.1   Install air heater sheet 1

Description	MH	Unit
Tube sheets	28.00	Ton
Baffle sheets—turn and vibration	60.00	Ton
Supports—tube sheets	20.00	Ton
Erect grid support steel	28.00	Ton
Supports—main	18.00	Ton
Install tubes	1.65	EA
Expand 2″ ends	1.15	EA
Expand 2-1/2″ ends	1.27	EA
Casing, veneer including batten bars and channels	64.00	Ton
Casing, main and hoppers including expansion joints and doors	60.00	Ton
Dampers	40.00	Ton

8.17   Air heater equipment installation estimate

8.17.1   Install air heater sheet 1

	Installation man-hour				
	Historical		Estimate		
	MH	Unit	Qty	Unit	BM
Install Air Heater Sheet 1					0
Tube sheets	28.00	Ton	0	Ton	0
Baffle sheets—turn and vibration	60.00	Ton	0	Ton	0
Supports—tube sheets	20.00	Ton	0	Ton	0
Erect grid support steel	28.00	Ton	0	Ton	0
Supports—main	18.00	Ton	0	Ton	0
Install tubes	1.65	EA	0	EA	0
Expand 2″ ends	1.15	EA	0	EA	0
Expand 2-1/2″ ends	1.27	EA	0	EA	0
Casing, veneer including batten bars and channels	64.00	Ton	0	Ton	0
Casing, main and hoppers including expansion joints and doors	60.00	Ton	0	Ton	0
Dampers	40.00	Ton	0	Ton	0

8.18 Air heater installation man hours

	Estimated	Actual
	BM	BM
Tube sheets	0	0
Air heater installation man-hours	0	0

8.19 General scope of field work required for each selective catalytic reduction unit

SCR reactor
Assemble shell
Field joint
Ammonia grid module
Field joint
Remove shipping angles
Install tie plate
Seal frame
Catalyst modules
Install catalyst

8.20 Selective catalytic reduction unit estimate data

8.20.1 Install selective catalytic reduction unit sheet 1

Description	MH	Unit
Assemble shell; set SCR reactor	20.00	Ton
Field joint	0.40	LF
Ammonia grid module	20.00	Ton
Remove shipping angles	40.00	Module
Install tie plate	40.00	Module
Seal frame	20.00	Ton
Catalyst modules	10.00	EA
Install catalyst	4.00	EA

8.21 Selective catalytic reduction unit equipment installation estimate

8.21.1 Install selective catalytic reduction unit sheet 1

	Installation man-hour				
	Historical			Estimate	
	MH	Unit	Qty	Unit	BM
Install selective catalytic reduction unit					0
SCR reactor	20.00	Ton		Ton	0
Assemble shell	20.00	Ton		Ton	0
Field joint	0.40	LF		LF	0
Ammonia grid module	20.00	Ton		Ton	0
Field joint	0.40	LF		LF	0
Remove shipping angles	40.00	Module		Module	0
Install tie plate	40.00	Module		Module	0
Seal frame	20.00	Ton		Ton	0
Catalyst modules	10.00	EA		Ton	0
Install catalyst	4.00	EA		EA	0

8.22 Selective Catalytic Reduction Unit Installation Man Hours

	Estimated	Actual
	BM	BM
Install selective catalytic reduction unit	0	0
Selective catalytic reduction unit installation man-hours	0	0

8.23 General scope of field work required for each FD, ID, primary air fan & pulverized seal air fan

Fans
Perform all millwright work for the four fans. Alignment and grouting
Install sole and base plates
Attach fan to foundation—bolted connection

FD Fan
Set fan housing split
Set impeller assembly/shaft
Alignment fan bearings
Seal weld housing inside/outside
Weld inlet cones and stud rings
Weld shaft seal stud rings
Coupling alignment
Install damper

ID Fan
Set fan housing split
Set impeller assembly/shaft
Alignment fan bearings
Seal weld housing inside/outside
Weld inlet cones and stud rings
Weld shaft seal stud rings
Coupling alignment
Install damper

Primary Air Fan
Set fan housing split
Set impeller assembly/shaft
Alignment fan bearings
Seal weld housing inside/outside
Weld inlet cones and stud rings
Weld shaft seal stud rings
Coupling alignment
Install damper

Pulverizer Seal Air Fan
Set fan housing split
Set impeller assembly/shaft
Alignment fan bearings
Seal weld housing inside/outside
Weld inlet cones and stud rings
Weld shaft seal stud rings
Coupling alignment
Install damper

8.24   FD, ID, primary air fan & pulverizer seal air fan estimate data

8.24.1   Install FD, ID, primary air fan & pulverizer seal air fan sheet 1

Description	MH	Unit
Install sole and base plates	2.00	EA
Attach fan to foundation—bolted connection	0.42	EA
Set fan housing split	4.00	EA
Set impeller assembly/shaft	4.00	EA
Alignment fan bearings	10.00	EA
Seal weld housing inside/outside	12.00	Job
Weld inlet cones and stud rings	4.00	Job
Weld shaft seal stud rings	4.00	Job
Coupling alignment	6.00	Job
Install damper	10.00	EA

8.25 FD, ID, primary air fan & pulverizer seal air fan equipment installation estimate

8.25.1 Install FD, ID, primary air fan & pulverizer seal air fan sheet 1

	Historical		Estimate		
	MH	Unit	Qty	Unit	BM
Fans					0
Install sole and base plates	2.00	EA		EA	0
Attach fan to foundation—bolted connection	0.42	EA		EA	0
FD Fan					0
Set fan housing split	4.00	EA		EA	0
Set impeller assembly/shaft	4.00	EA		EA	0
Alignment fan bearings	10.00	EA		EA	0
Seal weld housing inside/outside	12.00	Job		Job	0
Weld inlet cones and stud rings	4.00	Job		Job	0
Weld shaft seal stud rings	4.00	Job		Job	0
Coupling alignment	6.00	Job		Job	0
Install damper	10.00	EA		EA	0
ID Fan					0
Set fan housing split	4.00	EA		EA	0
Set impeller assembly/shaft	4.00	EA		EA	0
Alignment fan bearings	10.00	EA		EA	0
Seal weld housing inside/outside	12.00	Job		Job	0
Weld inlet cones and stud rings	4.00	Job		Job	0
Weld shaft seal stud rings	4.00	Job		Job	0
Coupling alignment	6.00	Job		Job	0
Install damper	10.00	EA		EA	0
Primary Air Fan					0
Set fan housing split	4.00	EA		EA	0
Set impeller assembly/shaft	4.00	EA		EA	0
Alignment fan bearings	10.00	EA		EA	0
Seal weld housing inside/outside	12.00	Job		Job	0
Weld inlet cones and stud rings	4.00	Job		Job	0
Weld shaft seal stud rings	4.00	Job		Job	0
Coupling alignment	6.00	Job		Job	0
Install damper	10.00	EA		EA	0
Pulverizer Seal Air Fan					0
Set fan housing split	4.00	EA		EA	0
Set impeller assembly/shaft	4.00	EA		EA	0
Alignment fan bearings	10.00	EA		EA	0
Seal weld housing inside/outside	12.00	Job		Job	0
Weld inlet cones and stud rings	4.00	Job		Job	0
Weld shaft seal stud rings	4.00	Job		Job	0
Coupling alignment	6.00	Job		Job	0
Install damper	10.00	EA		EA	0

8.26   FD, ID, primary air fan & pulverized seal air fan installation man hours

	Estimated	Actual
	BM	BM
Fans	0	0
FD fan	0	0
ID fan	0	0
Primary air fan	0	0
Pulverizer seal air fan	0	0
FD, ID, primary air fan, and pulverized seal air fan installation man-hours	0	0

8.27   General scope of field work required for each bottom ash handling system

Equipment consists of refractory-lined bottom ash chute and one (1) submerged ash drag chain conveyor.

Bottom ash chute
Submerged ash drag conveyor with outlet chute
Shipped in sections:
Head section—motor and drive shipped attached to head
Neck section
Transition section
Straight section
Tail section—take up shipped assembled
Scope of work:
Assembly of housing
Assembly of drag chain
Installation of drive units
Pneumatic/hydraulic take-up units, including air piping
Discharge chutes
Grouting
Refractory-lined chute on inlet of submerged ash drag
All water piping to and from submerged ash drag and drains

8.28 Bottom ash handling system estimate data

8.28.1 Install bottom ash handling system sheet 1

Description	MH	Unit
Bottom ash drag system		
Drive, head and neck section	120.00	Section
Refractor-lined ash chute	21.30	Ton
Grout	120.00	Lot
Water piping	120.00	LF

8.29 Bottom ash handling system installation estimate

8.29.1 Install bottom ash handling system sheet 1

	Installation man-hour				
	Historical			Estimate	
	MH	Unit	Qty	Unit	BM
Bottom ash drag system					0
Drive, head and neck section; lot 7,000 lb	120.00	Section	0.0	Section	0
Transition and trough section; lot 12,000 lb	120.00	Section	0.0	Section	0
Trail and trough section; lot 12,000 lb	120.00	Section	0.0	Section	0
Refractor-lined ash chute; lot 15,000 lb	21.30	TON	0.0	TON	0
Grout	120.00	Lot	0.0	Lot	0
Water piping	120.00	LF	0.0	LF	0

8.30 Bottom ash handling system—Installation man hours

	Estimated	Actual
	BM	BM
Bottom ash drag system	0	0
Bottom ash handling system—installation man-hours	**0**	**0**

8.31 General scope of field work required ductwork and breeching

Install ductwork and breeching and expansion joints. Fit up and seal welding from inside.
   Fit up flanges tack welded to sections to install and bolt field joints
Ductwork will rest on sliding supports or hang from supports

8.32   ID, FD, primary air and over fire air ductwork estimate data

8.32.1   Install ID, FD & primary air ductwork sheet 2

Description	MH	Unit
ID, FD, and primary air ductwork		
Duct transition, reducer, elbow, and sections	32.00	Ton
Bolted connection	0.26	EA
Field joint—weld flange	0.35	LF
Expansion joint	40.00	Ton
Duct support	40.00	Ton

8.32.2   Over fire ductwork sheet 2

Description	MH	Unit
Over fire ductwork		
Duct transition, reducer, elbow, and sections	32.00	Ton
Bolted connection	0.26	EA
Field joint—weld flange	0.35	LF
Expansion joint	40.00	Ton
Duct support	40.00	Ton

8.33 ID, FD, primary air and over fire air ductwork installation estimate

8.33.1 Install ID, FD & primary air ductwork sheet 1

	Historical			Estimate	
	MH	Qty	Unit	BM	BM
ID ductwork					**0**
Transition duct section	32.00	Ton	0	Ton	0
Bolted connection-duct transition to ID fan	0.26	EA	0	EA	0
Duct sections	32.00	Ton	0	Ton	0
Duct elbow	32.00	Ton	0	Ton	0
Field joint—weld flange	0.35	LF	0	LF	0
Bolted connection-duct sections	0.26	EA	0	EA	0
Expansion joint	40.00	Ton	0	Ton	0
Duct support	40.00	Ton	0	Ton	0
Primary air ductwork					**0**
Transition duct section	32.00	Ton	0	Ton	0
Bolted connection-duct transition to ID fan	0.26	EA	0	EA	0
Duct sections	32.00	Ton	0	Ton	0
Duct elbow	32.00	Ton	0	Ton	0
Field joint—weld flange	0.35	LF	0	LF	0
Bolted connection-duct sections	0.26	EA	0	EA	0
Expansion joint	40.00	Ton	0	Ton	0
Duct support	40.00	Ton	0	Ton	0
FD ductwork					**0**
Transition duct section	32.00	Ton	0	Ton	0
Bolted connection-duct transition to ID fan	0.26	EA	0	EA	0
Duct sections	32.00	Ton	0	Ton	0
Duct elbow	32.00	Ton	0	Ton	0
Field joint—weld flange	0.35	LF	0	LF	0
Bolted connection-duct sections	0.26	EA	0	EA	0
Expansion joint	40.00	Ton	0	Ton	0
Duct support	40.00	Ton	0	Ton	0

8.33.2 Install over fire ductwork sheet 2

	Installation man-hour				
	Historical			Estimate	
	MH	Qty	Unit	BM	BM
Over fire air duct					**0**
Transition duct section	32.00	Ton	0	Ton	0
Bolted connection-duct transition to ID fan	0.26	EA	0	EA	0
Duct sections	32.00	Ton	0	Ton	0
Duct elbow	32.00	Ton	0	Ton	0
Field joint—weld flange	0.35	LF	0	LF	0
Bolted connection-duct sections	0.26	EA	0	EA	0
Expansion joint	40.00	Ton	0	Ton	0
Duct support	40.00	Ton	0	Ton	0

8.34 ID, FD, primary air and over fire ductwork—Installation man hours

	Estimated	Actual
	BM	BM
ID ductwork	0	0
Primary air ductwork	0	0
FD ductwork	0	0
Over fire air duct	0	0
ID, OFA, FD, and over fire ductwork—installation man-hours	0	0

8.35 General scope of field work required for each electrostatic precipitator

Casing
Prefabricate wall and partition panels

Install anvil beams and antisneak age baffles on end frames/roof beams and bottom baffles

Set partition and wall panels; stabilize with top-end frames, bottom-end frames and intermittently with intermediate roof beams and supporting bottom baffles

Install remaining intermediate roof beams and supporting bottom baffles

Square and plumb casing and began weld out

Complete casing welding, including the hoppers

Drop in preassembled hoppers with lower alignment frames and scaffolding inside

Secure interior cross ties between roof beams, end frames, and bottom baffles

Hang inlet and outlet perforated plates

Preassemble and mount inlet and outlet plenums

Install inlet and outlet plenum cone sections

Install platforms, stairways, and ladders

Internals
Lift and set collecting plates into ESP

Join plate halves and attach plates to anvil beams

Do preliminary alignment of collecting plates

Preassemble discharge electrodes to high voltage frames

Lift high voltage frames and rest on collecting plates

Penthouse
Preassemble hot roof sections

Install hot roof sections

Install lower alignment frames to discharge electrodes

Suspend high voltage frames from support insulators

Complete welding of penthouse walls and hot roof

Weld out rapper sleeves (installed below)

ESP roof
Preassemble cold roof sections

Insulate underside of cold roof

Set cold roof sections and expansion joints

Install roof handrail

Install and weld out rapper sleeves

Weld out cold roof

Install rapper shafts, rapper adjusting studs, and rappers

Align rappers

Set TR sets

Install duct support steel
Install inlet ductwork on the support steel including the louver dampers
Install seal air system platform on duct support steel
Set seal air fan and heater
Support structure
Sliding plates and cap plates

8.36 Electrostatic precipitator estimate data

8.36.1 Hoppers, casing/partition frames, support steel, diffusers, TEF's & IRB's sheet 1

Description	MH	Unit
Preassemble hoppers		
Place hopper cones and baffles	1.80	0
Place side wall sections	1.50	0
Weld side wall and hopper baffles	0.32	LF
Clean out door	4.00	EA
Preassemble casing and partition frames		
Place and fit side and partition frame sections	10.00	0
Weld side frame and partition sections—3/8″ square butt weld	0.75	LF
Bevel beams 1/2″ thick × 12″	2.50	Ea.
Place and fit beams	10.00	Ea.
Full penetration field weld 1/2″ double bevel	16.80	Ea.
Erect precipitator support steel		
Erect and bolt structural columns	40.00	Ea.
Erect X—bracing	21.00	Ea.
Bolt X—bracing	3.40	Ea.
Preassemble diffusers	240.00	Job
Preassemble perforated plates		
Load, haul, offload inlet/outlet perforated plates 11′-3″ × 5′-0″	2.00	Ea.
Place and fit perforated plate sections	4.00	Ea.
Weld perforate plate sections	0.35	LF
Place and weld clips	1.00	Ea.
Preassemble TEFs and IRBs		
Load, haul, offload frames	2.76	Ton
Set top-end frames	20.00	Ea.
Set intermediate roof beams	32.00	Ea.

8.36.2   Slide plates, hopper platform, support steel access sheet 2

Description	MH	Unit
Set precipitator slide plates		
Set slide plates	4.00	Ea.
Erect hopper platform		
Erect platform steel	0.10	SF
Erect grating	0.15	SF
Erect handrail/toe plate	0.25	LF
Install support steel access		
Erect platform steel	0.29	SF
Erect grating	0.24	SF
Erect handrail/toe plate	0.36	LF
Stairs	0.85	LF
Ladder	0.30	LF

8.36.3   Penthouse, casing, hoppers, weld casing, frames/beams sheet 3

Description	MH	Unit
Preassemble penthouse		
Place and fit penthouse roof and side frame sections	2.50	Ton
Weld penthouse roof sections and sides/ends	2.15	LF
Erect casing		
Rig and place side and partition frame casing	40.00	Ea.
Attach temporary guy lines to casing/turn buckles	6.67	Ea.
Set hoppers		
Rig and place hoppers on support steel	1.50	Ton
Field bolt list	0.15	EA
Plumb, square, weld casing		
	360.00	Job
Weld side frame and partition sections—3/8″ square butt weld	0.75	LF
Weld side wall casing corners	0.35	LF
Weld outside side wall casing 2–10	0.35	LF
Bevel beams 1/2″ thick × 12″	2.50	Ea.
Place and fit beams	10.00	Ea.
Full penetration field weld 1/2″ double bevel	16.00	Ea.
Field bolts	300.00	Lot
Install top-end frames and intermediate roof beams		
Place and fit top-end frames/roof beams	40.00	Ton
Field bolts	0.20	Ea.
Seal weld around bolts/field weld 1/4″ fillet	0.35	LF
Full penetration field weld 1/2″ double bevel	16.00	Ea.

8.36.4 Frames/baffles, collecting plates, hopper auxiliaries, electrodes, hot roof sheet 4

Description	MH	Unit
Install bottom-end frames and bottom support baffles		
Place and fit bottom-end frames and baffles	2.50	Ton
Field bolts	0.20	Ea.
Seal weld around bolts/field weld 1/4″ fillet	0.35	LF
Full penetration field weld 1/2″ double bevel	16.00	Ea.
Install collecting plates		
Install collecting plates	3.50	Ea.
Bolt plates	1.00	Ea.
Weld corner	240.00	Job
Decking	160.00	Job
Install hopper auxiliaries		
Install pneumatic hopper slide gates (12″ × 12″)	7.80	Assembly
Install manual hopper slide gates (12″ × 12″)	7.80	Assembly
Install TR controllers	6.00	Assembly
Install upper frames/electrodes		
Install high voltage frames	16.00	Ea.
Load, haul, offload discharge electrodes	0.50	Ea.
Install spiked discharge electrodes 2″ diameter × 40′	1.20	Ea.
Preliminary alignment	480.00	Job
Install/weld hot roof		
Place and fit casing roof sections	27.50	Ton
Weld roof casing	0.35	LF
Weld roof casing at corners	0.35	LF

8.36.5 Set plates, rapper rods, Mfg. plates, penthouse, diffuser, alignment, rectifiers sheet 5

Description	MH	Unit
Set precipitator perforated plates		
Set perforated plate sections	10.00	Ea.
Weld perforated plate sections	0.35	LF
Place and weld clips	1.50	Ea.
Set rapper rods		
Rapper sleeves, shafts and insulators, DE alumina rapper shafts	1.10	Ea.
Seal weld sleeves	0.35	Ea.
Set support bushings/mfg. plates		
Set and weld, bolt mfg. plates	1.00	Ea.
Set support bushings	0.24	Ea.
Erect penthouse		
Place and fit penthouse roof sections, sides, and ends	2.5	Ton
Weld penthouse roof, sides, ends, and corners	2.15	LF
Erect and weld in/out diffusers	1.00	Ea.
Final alignment	2.10	Ton
Set transformer rectifier sets		
Load, haul, offload transformer rectifiers	4.00	Ea.
Set transformer rectifiers	20.00	Ea.
Install rappers	3.00	Ea.
Install rapper controllers	20.00	Ea.

8.36.6 Seal air system, dampers, stack, inlet/outlet plenum sheet 6

Description	MH	Unit
Erect seal air system		
Install seal air blower and heater system	260.00	Ea.
Install dampers and access doors		
Install dampers	50.00	Ea.
Install 30″ × 30″ access doors	4.09	Ea.
Erect stack		
Set bottom and upper section	60.00	Ea.
Bolt bottom section to foundation	1.53	Ea.
Fit and weld upper sections	0.75	LF
Platform—360 deg	0.10	SF
Grating	0.15	SF
Handrail	0.25	elf
Step platforms	6.00	Ea.
Caged ladder	0.35	LF
Preassemble and erect inlet/outlet plenum		
Place and fit inlet and outlet plenum sections	11.40	Ton
Weld inlet and outlet plenum sections	0.75	LF
Erect inlet/outlet plenum		
Place and fit inlet plenum sections	22.00	Ton
Place and fit outlet plenum sections	40.00	Ton
Weld inlet and outlet plenum sections	0.75	LF

8.36.7 Penthouse structural steel sheet 7

Description	MH	Unit
Penthouse structural steel and platforms		
Erect structural steel	8.00	pcs
Bolted connections	1.00	Ea.
Platforms and stairs	160.00	lot

8.37   Electrostatic precipitator installation estimate

8.37.1   Hoppers, casing/partition frames, erect steel, diffusers, TEF/IRB sheet 1

Description	Historical		Estimate		
	MH	Unit	Qty	Unit	BM
Preassemble hoppers					0
Set hopper cones	1.80	Ton	0	Ton	0
Place and fit hopper side wall sections	1.50	Ton	0	Ton	0
Weld side wall sections	0.32	LF	0	LF	0
Place and fit hopper baffle	1.80	Ton	0	Ton	0
Weld hopper baffle 3/16" fillet/1/8" fillet 2–10	0.32	LF	0	LF	0
Clean out door	4.00	EA	0	EA	0
Preassemble casing and partition frames					0
Place and fit side frame sections	10.00	Ton	0	Ton	0
Weld and fit side frame sections—3/8" square butt weld	0.75	LF	0	LF	0
Place and fit Partition frame sections	10.00	Ton	0	Ton	0
Weld and fit partition frame sections—3/8" square butt weld	0.75	LF	0	LF	0
Bevel beams 1/2" thick × 12"	2.50	Ea.	0	Ea.	0
Place and fit beams	10.00	Ea.	0	Ea.	0
Full penetration field weld 1/2" double bevel	16.80	Ea.	0	Ea.	0
Erect precipitator support steel					0
Erect and bolt structural columns	40.00	Ea.	0	Ea.	0
Erect X—bracing	21.00	Ea.	0	Ea.	0
Bolt X—bracing	3.40	Ea.	0	Ea.	0
Preassemble diffusers	240.00	Job	0	Job	0
Preassemble perforated plates					0
Load, haul, offload inlet/outlet perforated plates	2.00	Ea.	0	Ea.	0
Place and fit perforated plate sections	4.00	Ea.	0	Ea.	0
Weld perforated plate sections	0.35	LF	0	LF	0
Place and weld clips	1.00	Ea.	0	Ea.	0
Preassemble TEFs and IRBs					0
Load, haul, offload frames	2.76	Ton	0	Ton	0
Set top-end frames	20.00	Ea.	0	Ea.	0
Set intermediate roof beams	32.00	Ea.	0	Ea.	0

8.37.2 Slide plates, hopper platform, support steel access sheet 2

Description	Historical		Qty	Estimate	
	MH	Unit		Unit	BM
Set precipitator slide plates					**0**
Set slide plates	4.00	Ea.	0	Ea.	0
Erect hopper platform					**0**
Erect platform steel	0.10	SF	0	SF	0
Erect grating	0.15	SF	0	SF	0
Erect handrail	0.25	LF	0	LF	0
Install support steel access					**0**
Stair towers from side access level to top ESP roof					
Erect platform steel	0.29	SF	0	SF	0
Erect grating	0.24	SF	0	SF	0
Erect handrail/toe plate	0.36	LF	0	LF	0
Stairs	0.85	Ea.	0	Ea.	0
Stair towers from grade to side access level					
Erect platform steel	0.29	SF	0	SF	0
Erect grating	0.24	SF	0	SF	0
Erect handrail/toe plate	0.36	LF	0	LF	0
Stairs	0.85	Ea.	0	Ea.	0
Side access level					
Erect platform steel	0.29	SF	0	SF	0
Erect grating	0.24	SF	0	SF	0
Erect handrail/toe plate	0.36	LF	0	LF	0
Ladder	0.30	LF	0	LF	0
Caged ladders from grade to roof					
Erect platform steel	0.29	SF	0	SF	0
Erect grating	0.24	SF	0	SF	0
Erect handrail/toe plate	0.36	LF	0	LF	0
Ladder	0.30	LF	0	LF	0
Side access to hopper platform					
Erect platform steel	0.29	SF	0	SF	0
Erect grating	0.24	SF	0	SF	0
Erect handrail/toe plate	0.36	LF	0	LF	0
Handrail on ESP roof perimeter					
Erect handrail/toe plate	0.36	LF	0	LF	0

8.37.3 Penthouse, casing, hoppers, weld casing, frames/beams sheet 3

Description	Historical			Estimate	
	MH	Unit	Qty	Unit	BM
Preassemble penthouse					**0**
Place and fit penthouse roof sections	2.50	Ton	0	Ton	0
Weld penthouse roof sections	2.15	LF	0	LF	0
Place and fit penthouse sides and ends	2.50	Ton	0	Ton	0
Weld penthouse sides and ends	2.15	LF	0	LF	0
Erect casing					**0**
Rig and place side frame casing (row 1 and 3)	40.00	Ea.	0	Ea.	0
Rig and place partition frame casing (row 2)	40.00	Ea.	0	Ea.	0
Attach temporary guy lines to casing/turn buckles	6.67	Ea.	0	Ea.	0
Set hoppers					**0**
Rig and place hoppers on support steel	1.50	Ton	0	Ton	0
Field bolt list	0.15	EA	0	EA	0
Plumb, square, weld casing					**0**
Plumb, square, and fit casing (turnbuckle)	360.00	Job	0	Job	0
Weld side frame sections—3/8″ square butt weld	0.75	LF	0	LF	0
Weld partition frame sections—3/8″ square butt weld	0.75	LF	0	LF	0
Weld side wall casing corners	0.35	LF	0	LF	0
Weld outside side wall casing 2–10	0.35	LF	0	LF	0
Bevel beams 1/2″ thick × 12″	2.50	Ea.	0	Ea.	0
Place and fit beams	10.00	Ea.	0	Ea.	0
Full penetration field weld 1/2″ double bevel	16.00	Ea.	0	Ea.	0
Field bolts	300.00	Lot	0	Lot	0
Install top-end frames and intermediate roof beams					**0**
Place and fit top-end frames	40.00	Ton	0	Ton	0
Place and fit roof beams	40.00	Ton	0	Ton	0
Field bolts	0.20	Ea.	0	Ea.	0
Seal weld around bolts	0.35	Ea.	0	Ea.	0
Field weld 1/4″ fillet	0.35	LF	0	LF	0
Full penetration field weld 1/2″ double bevel	16.00	Ea.	0	Ea.	0

8.37.4   Frames/baffles, collecting plates, hopper auxiliaries, electrodes, hot roof sheet 4

Description	Historical			Estimate	
	MH	Unit	Qty	Unit	BM
Install bottom-end frames and bottom support baffles					0
Place and fit bottom-end frames	2.50	Ton	0	Ton	0
Place and fit baffles	2.50	Ton	0	Ton	0
Field bolts	0.20	Ea.	0	Ea.	0
Seal weld around bolts	0.35	LF	0	LF	0
Field weld 1/4″ fillet	0.35	LF	0	LF	0
Full penetration field weld 1/2″ double bevel	16.00	Ea.	0	Ea.	0
Install collecting plates					0
Install collecting plates	3.50	Ea.	0	Ea.	0
Bolt plates	1.00	Ea.	0	Ea.	0
Weld corner	240.00	Job	0	Job	0
Decking	160.00	Job	0	Job	0
					0
Install pneumatic hopper slide gates (12″ × 12″)	7.80	Assembly	0	Assembly	0
Install manual hopper slide gates (12″ × 12″)	7.80	Assembly	0	Assembly	0
Install TR controllers	6.00	Assembly	0	Assembly	0
Install upper frames/electrodes					0
Install high voltage frames	16.00	Ea.	0	Ea.	0
Load, haul, offload discharge electrodes	0.50	Ea.	0	Ea.	0
Install spiked discharge electrodes 2″ diameter × 40′	1.20	Ea.	0	Ea.	0
Preliminary alignment	480.00	Job	0	Job	0
Install/weld hot roof					0
Place and fit casing roof sections	27.50	Ton	0	Ton	0
Weld roof casing	0.35	LF	0	LF	0
Weld roof casing at corners	0.35	LF	0	LF	0

8.37.5 Perforated plates, rapper rods, Mfg. plates, penthouse, diffuser, alignment, rectifier set sheet 5

Description	Historical		Estimate		
	MH	Unit	Qty	Unit	BM
Set precipitator perforated plates					**0**
Set perforated plate sections	4.00	Ea.	0	Ea.	0
Weld perforated plate sections	0.35	LF	0	LF	0
Place and weld clips	1.00	Ea.	0	Ea.	0
Set rapper rods					**0**
Set plate rapper sleeves 4″ diameter × 10′	1.10	Ea.	0	Ea.	0
Seal weld sleeves	0.35	Ea.	0	Ea.	0
Set plate rapper shafts 2″ diameter x 10′	1.10	Ea.	0	Ea.	0
Set alumina support insulators	1.10	Ea.	0	Ea.	0
Set DE alumina rapper shafts	1.10	Ea.	0	Ea.	0
Set support bushings/mfg. plates					**0**
Set mfg. plates	1.00	Ea.	0	Ea.	0
Bolt and weld mfg. plates	1.00	Ea.	0	Ea.	0
Set support bushings	0.24	Ea.	0	Ea.	0
Erect penthouse					**0**
Place and fit penthouse roof sections	2.5	Ton	0	Ton	0
Weld penthouse roof sections	2.15	LF	0	LF	0
Place and fit penthouse sides and ends	2.5	Ton	0	Ton	0
Weld penthouse sides and ends	2.15	LF	0	LF	0
Weld penthouse sides and ends at corners	2.15	LF	0	LF	0
Erect and weld in/out diffusers	1.00	Ea.	0	Ea.	0
Final alignment	2.10	Ton	0	Job	0
Set transformer rectifier sets					**0**
Load, haul, offload transformer rectifiers	4.00	Ea.	0	Ea.	0
Set transformer rectifiers	20.00	Ea.	0	Ea.	0
Install rappers					**0**
Install rappers	3.00	Ea.	0	Ea.	0
Install rapper controllers	20.00	Ea.	0	Ea.	0

8.37.6   Seal air system, dampers, stack, inlet/outlet plenum sheet 6

Description	MH	Historical Unit	Qty	Estimate Unit	BM
Erect seal air system					**0**
Install seal air blower and heater system	260.00	Ea.	0	Ea.	0
Install dampers and access doors					**0**
Install dampers	50.00	Ea.	0	Ea.	0
Install 30″ × 30″ access doors	4.00	Ea.	0	Ea.	0
Erect stack					**0**
Set bottom section	60.00	Ea.	0	Ea.	0
Bolt bottom section to foundation	1.53	Ea.	0	Ea.	0
Set upper sections	60.00	Ea.	0	Ea.	0
Fit and weld upper sections	0.75	LF	0	LF	0
Install platforms at ground level					
Platform—360 deg	0.10	SF	0	SF	0
Grating	0.15	SF	0	SF	0
Handrail	0.25	LF	0	LF	0
Step platforms	6.00	Ea.	0	Ea.	0
Caged ladder	0.35	LF	0	LF	0
Preassemble inlet/outlet plenum					**0**
Place and fit inlet plenum sections	11.40	Ton	0	Ton	0
Weld inlet plenum sections	0.75	LF	0	LF	0
Place and fit outlet plenum sections	11.40	Ton	0	Ton	0
Weld outlet plenum sections	0.75	LF	0	LF	0
Erect inlet/outlet plenum					**0**
Place and fit inlet plenum sections	22.00	Ton	0	Ton	0
Weld inlet plenum sections	0.75	LF	0	LF	0
Place and fit outlet plenum sections	40.00	Ton	0	Ton	0
Weld outlet plenum sections	0.75	LF	0	LF	0

8.37.7   Penthouse structural steel sheet 7

Description	MH	Historical Unit	Qty	Estimate Unit	BM
Penthouse structural steel and platforms					**0**
Erect structural steel	8.00	pcs	0	pcs	0
Bolted connections	1.00	Ea.	0	Ea.	0
Platforms and stairs	160.00	Lot	0	Lot	0

8.38 Electrostatic precipitator-installation man-hours

Scope of work	Estimated BM	Actual BM
Preassemble hoppers	0	0
Preassemble casing and partition frames	0	0
Erect precipitator support steel	0	0
Preassemble diffusers	0	0
Preassemble perforated plates	0	0
Preassemble TEFs and IRBs	0	0
Set precipitator slide plates	0	0
Erect hopper platform	0	0
Install support steel access	0	0
Preassemble penthouse	0	0
Erect casing	0	0
Set hoppers	0	0
Plumb, square, weld casing	0	0
Install top-end frames and intermediate roof beams	0	0
Install bottom-end frames and bottom support baffles	0	0
Install collecting plates	0	0
Install hopper auxiliaries	0	0
Install upper frames/electrodes	0	0
Preliminary alignment	0	0
Install/weld hot roof	0	0
Set precipitator perforated plates	0	0
Set rapper rods	0	0
Set support bushings/mfg. plates	0	0
Erect penthouse	0	0
Erect and weld in/out diffusers	0	0
Final alignment	0	0
Set transformer rectifier sets	0	0
Install rappers	0	0
Erect seal air system	0	0
Install dampers and access doors	0	0
Erect stack	0	0
Preassemble inlet/outlet plenum	0	0
Erect inlet/outlet plenum	0	0
Penthouse structural steel and platforms	0	0
Electrostatic precipitator installation man-hours	**0**	**0**

8.39   Scope of field work required for structural steel and boiler casing

Structural Steel
Erect Boiler Structural Steel (Stairs, Platforms, Grating, and Handrails)
Main steel
Platform framing
Grating
Handrail
Erect Penthouse Steel
Install Platform, Grating, and Handrail
Platform framing
Grating
Handrail
Install Stair Nos. 1 and 2
Structural steel
Platform framing
Grating
Handrail
Stair treads
Erect Casing—Generating Bank, Boiler, and Penthouse

8.40   Structural steel and boiler casing estimate data

8.40.1   Structural steel and boiler casing sheet 1

Description	MH	Unit
Erect boiler structural steel (stairs, platforms, grating, and handrails)		
Main steel	24.00	Ton
Platform framing	0.15	SF
Grating	0.20	SF
Handrail	0.25	LF
Erect penthouse steel	24.00	Ton
Install platform, grating, and handrail at El. 100′-0″ to 184′0″		
Platform framing	0.15	SF
Grating	0.20	SF
Handrail	0.25	LF
Install stair nos. 1 and 2		
Structural steel	24.00	Ton
Platform framing	0.15	SF
Grating	0.20	SF
Handrail	0.25	LF
Stair treads	0.85	EA
Flues and ducts—with supports, expansion joints, and dampers	40.00	Ton
Coupling guards	14.00	Ton
Erect casing—generating bank, boiler, and penthouse	62.00	Ton

8.41   Structural steel and boiler casing installation estimate

8.41.1   Erect structural steel and boiler casing sheet 1

	Installation man-hour				
	Historical			Estimate	
	MH	Unit	Qty	BM	BM
Flues and ducts					**0**
Flues and ducts	40.00	Ton		Ton	0
Coupling guards	14.00	Ton		Ton	0
Structural steel					
Erect boiler structural steel					**0**
Main steel	24.00	Ton		Ton	0
Platform framing	0.15	SF		SF	0
Grating	0.20	SF		SF	0
Handrail	0.25	LF		LF	0
Erect penthouse steel, platform, grading, and handrail					**0**
Main steel	24.00	Ton		Ton	0
Platform framing	0.15	SF		SF	0
Grating	0.20	SF		SF	0
Handrail	0.25	LF		LF	0
Install stair nos. 1 and 2					**0**
Structural steel	24.00	Ton		Ton	0
Platform framing	0.15	SF		SF	0
Grating	0.20	SF		SF	0
Handrail	0.25	LF		LF	0
Stair treads	0.85	EA		EA	0
Erect casing—generating bank, boiler, and penthouse	62.00	Ton		Ton	**0**

8.42   Structural steel and boiler installation man-hours estimated actual

	BM	BM
Flues and ducts	0	0
Erect boiler structural steel	0	0
Erect penthouse steel, platform, grading, and Handrail	0	0
Install stair nos. 1 and 2	0	0
Erect casing—generating bank, boiler, and penthouse	0	0
Structural steel and boiler casing installation man-hours	**0**	**0**

8.43 Coal-fired boiler and auxiliary equipment man-hour breakdown

	Actual	Estimate
	MH	BM
Coal-fired boiler installation	0	0
Coal silo and pulverized installation	0	0
Air heater installation	0	0
Selective catalytic reduction unit installation	0	0
FD, ID, primary air fan, and pulverized seal air fan installation	0	0
Bottom ash handling system—installation	0	0
ID, OFA, FD, and over fire ductwork—installation	0	0
Electrostatic precipitator installation	0	0
Structural steel and boiler casing installation	0	0
Coal-fired boiler and auxiliary equipment man-hour breakdown	**0**	**0**

Chapter 9

Solar thermal power plant

9.1 Solar thermal power plant

Solar thermal power systems use concentrated energy. Solar thermal power (electricity) generation systems collect and concentrate sunlight to produce high temperatures needed to generate electricity. All solar thermal power systems have solar energy collectors with two main components: reflectors (mirrors) that capture and focus sunlight onto a receiver and a heat transfer fluid heated and circulated in the receiver and used to produce steam. The steam is converted into energy in a turbine, which powers a generator to produce electricity.

Types of solar power plants
Linear concentrating systems, which include parabolic roughs and linear Fresnel reflectors
Solar power towers

Linear concentrating systems
Linear concentrating systems collect the sun's energy using long, rectangular, curved (U-shaped) mirrors. The mirrors focus sunlight onto receivers (tubes) that run the length of the mirrors. The concentrated sunlight heats a fluid flowing through the tubes. The fluid is sent to heat exchanger to boil water in a steam-turbine generator to produce electricity. A linear concentrating collector power plant has a large number or field of collectors in parallel rows that are aligned in a north-south orientation to maximize solar energy collection. This configuration enables the mirrors to track the sun from east to west during the day and concentrate sunlight continuously onto the receiver tubes.

Solar power towers
A solar power system uses a large field of flat, sun-tracking mirrors called heliostats to reflect and concentrate sunlight onto a receiver on the top of a tower. Some power towers use water as the heat transfer fluid. Advanced designs use molten nitrate salt because of its superior heat transfer and energy storage capabilities.

Industrial Process Plant Construction Estimating and Man-Hour Analysis.
https://doi.org/10.1016/B978-0-12-818648-0.00009-0
187

9.2   General scope of field work required for each receiver

Scope of Work-Field Erection

Assemble L1, L2, and L3 core columns with crossbeams and cross braces

Lift and install deck crossbeams between columns

Lift L3 primary sun assembly with columns

Lift and install all secondary steel and install grating, handrails, and ladders

Lift and install all roof deck secondary steel and install grating, handrails, and ladders

Lift and seal weld crane base plate and roof deck diamond plates

Install crossbeams

Bolted connections—crossbeams

Lift ECV, load, haul, and set ECV

Connect vertical struts to ECV ring girder

Weld bumper plates to bumper stubs

EVC in final position

Set receiver sections on top of 600 ft tower

Connect L1, L2, and L3 core columns at splice

9.3   Receiver estimate data

9.3.1   Assemble and erect core columns and cross-braces sheet 1

Description	MH	Unit
Assemble columns; W10 × 77 × 39′	6.5	Ton
Install crossbeams; W10 × 49 × 10′	6.5	Ton
Install crossbeams; W16 × 40 × 10′	6.5	Ton
Install crossbeams; W10 × 50 × 10′	6.5	Ton
Install cross braces HSS8x8x.500	1.5	EA
Bolted connections—cross braces	1.5	EA
Lift and set	20	EA
Make up anchor bolts 8 ea.	0.5	EA

9.3.2   Install secondary steel, platforms and ladders sheet 2

Description	MH	Unit
W6x12	1.50	Ton
Bolted connections	1.00	EA
Install grating	0.20	SF
Install handrail	0.27	LF
Install caged ladder	0.35	LF
MC7x19.1	1.50	Ton
Bolted connections	1.00	EA
Weld mounting hardware	100.0	EA

9.3.3   Install secondary steel, vertical struts sheet 3

Description	MH	Unit
Vertical struts	20	EA
Lift ECV, load, haul, and set ECV	2.4	EA
Connect vertical struts to ECV ring girder	30	EA
Weld bumper plates to bumper stubs	30	EA

9.4   Receiver installation estimate

9.4.1   Assemble and erect L1 core columns and cross-braces sheet 1

		Historical		Estimate		
Description	MH	Qty	Unit	Qty	Unit	IW
Erect core columns L1						**0**
Assemble columns; W10 × 77 × 39′ (18 EA)	6.5	27.1	Ton	0	Ton	0.0
Install crossbeams; W10 × 49 × 10′ (46 EA)	6.5	11.3	Ton	0	Ton	0.0
Install crossbeams; W16 × 40 × 10′ (10 EA)	6.5	2.0	Ton	0	Ton	0.0
Install crossbeams; W10 × 50 × 10′ (41 EA)	6.5	10.3	Ton	0	Ton	0.0
Install cross braces HSS8x8x.500	1.5	175	EA	0	EA	0.0
Bolted connections—cross braces	1.5	602	EA	0	EA	0.0
Lift and set	20	10	EA	0	EA	0.0
Make up anchor bolts	0.5	160	EA	0	EA	0.0

9.4.2   Assemble and erect L2 core columns and cross-braces sheet 2

		Historical		Estimate		
Description	MH	Qty	Unit	Qty	Unit	IW
Erect core columns L2						**0**
Assemble columns; W10 × 77 × 39′ (18 EA)	6.5	54.2	Ton	0	Ton	0.0
Install crossbeams; W10 × 49 × 10′ (53 EA)	6.5	26.0	Ton	0	Ton	0.0
Install crossbeams; W16 × 40 × 10′ (6 EA)	6.5	2.4	Ton	0	Ton	0.0
Install crossbeams; W10 × 50 × 10′ (43 EA)	6.5	21.5	Ton	0	Ton	0.0
Install cross braces HSS8x8x.500	1.5	342	EA	0	EA	0.0
Bolted connections—cross braces	1.5	1508	EA	0	EA	0.0
Lift and set	20	22	EA	0	EA	0.0
Make up anchor bolts	0.5	288	EA	0	EA	0.0

9.4.3 Assemble and erect L3 core columns and cross-braces sheet 3

Description	MH	Historical Qty	Unit	Estimate Qty	Unit	IW
Erect core columns L3						**0**
Assemble columns; W10 × 77 × 39′ (18 EA)	6.5	54.2	Ton	0	Ton	0.0
Install crossbeams; W10 × 49 × 10′ (46 EA)	6.5	22.5	Ton	0	Ton	0.0
Install crossbeams; W16 × 40 × 10′ (9 EA)	6.5	3.6	Ton	0	Ton	0.0
Install crossbeams; W10 × 50 × 10′ (41 EA)	6.5	20.5	Ton	0	Ton	0.0
Install cross braces HSS8x8x.500	1.5	350	EA	0	EA	0.0
Bolted connections—cross braces	1.5	1016	EA	0	EA	0.0
Lift and set	20	20	EA	0	EA	0.0
Make up anchor bolts	0.5	320	EA	0	EA	0.0

9.4.4 Install L1 secondary steel, platforms sheet 4

Description	MH	Historical Qty	Unit	Estimate Qty	Unit	IW
Install L1 secondary steel						**0**
W6x12	1.50	56.0	Ton	0	Ton	0.0
Bolted connections	1.00	168.0	EA	0	EA	0.0
Install grating	0.20	2760.0	EA	0	EA	0.0
Install handrail	0.27	641.0	SF	0	SF	0.0
Install caged ladder	0.35	240.0	LF	0	LF	0.0
MC7x19.1	1.50	28.0	EA	0	EA	0.0
Bolted connections	1.00	56.0	EA	0	EA	0.0
Weld mounting hardware	100.0	1.0	EA	0	EA	0.0

9.4.5   Install L2 secondary steel, platforms sheet 5

	Historical			Estimate		
Description	MH	Qty	Unit	Qty	Unit	IW
Install L2 secondary steel						**0**
W6x12	1.50	56.0	Ton	0	Ton	0.0
Bolted connections	1.00	112.0	EA	0	EA	0.0
Install grating	0.20	1840.0	EA	0	EA	0.0
Install handrail	0.27	428.0	SF	0	SF	0.0
Install caged ladder	0.35	160.0	LF	0	LF	0.0
MC7x19.1	1.50	28.0	EA	0	EA	0.0
Bolted connections	1.00	56.0	EA	0	EA	0.0
Weld mounting hardware	100.0	1.0	EA	0	EA	0.0
Vertical struts	20	3.0	EA	0		0.0
Lift ECV, load, haul, and set ECV	80	1.0	EA	0	EA	0.0
Connect vertical struts to ECV ring girder	30	1.0	EA	0	EA	0.0
Weld bumper plates to bumper stubs	30	1.0	EA	0	EA	0.0

9.4.6   Install L3 secondary steel, platforms sheet 6

	Historical			Estimate		
Description	MH	Qty	Unit	Qty	Unit	IW
Install L3 secondary steel						**0**
W6x12	1.50	84.0	Ton	0	Ton	0.0
Bolted connections	1.00	168.0	EA	0	EA	0.0
Install grating	0.20	2760.0	EA	0	EA	0.0
Install handrail	0.27	641.0	SF	0	SF	0.0
Install caged ladder	0.35	240.0	LF	0	LF	0.0
MC7x19.1	1.50	28.0	EA	0	EA	0.0
Bolted connections	1.00	56.0	EA	0	EA	0.0
Weld mounting hardware	100.0	1.0	EA	0	EA	0.0

9.5 Receiver installation man hours

	Estimated	Actual
	IW	IW
Erect core columns L1	1775	0
Erect core columns L2 (2 EA)	4035	0
Erect core columns L3 (2 EA)	3264	0
Install L1 secondary steel	1259	0
Install L2 secondary steel	1134	0
Install L3 secondary steel	1301	0
Receiver installation man-hours	12,768	0

9.6 General scope of field work required for receiver panel assemblies

Receiver panel assemblies
Top adjust assembly
Bolt connection
Lower adjust assembly
Bolt connection
Shim plates
Linkage attachment
Bolt connection
Linkage assembly
Panel

9.7 Receiver panel assembly estimate data

9.7.1 Assemble and erect receiver panels sheet 1

Description	MH	Unit
Receiver panel assemblies		
Top adjust assembly	8	EA
Bolt connection	0.25	EA
Lower adjust assembly	8	EA
Bolt connection	0.25	EA
Shim plates	2	EA
Linkage attachment	4	EA
Bolt connection	0.25	EA
Linkage assembly	10	EA
Panel	80	EA

9.8 Receiver panel assembly installation estimate

9.8.1 Assemble and erect receiver panels sheet 1

Description	Historical			Estimate		
	MH	Qty	Unit	Qty	Unit	IW
Receiver panel assemblies						**0**
Top adjust assembly	8	14	EA	0	EA	0
Bolt connection	0.25	84	EA	0	EA	0
Lower adjust assembly	8	14	EA	0	EA	0
Bolt connection	0.25	84	EA	0	EA	0
Shim plates	2	14	EA	0	EA	0
Linkage attachment	4	14	EA	0	EA	0
Bolt connection	0.25	56	EA	0	EA	0
Linkage assembly	10	14	EA	0	EA	0
Panel	80	14	EA	0	EA	0

9.9 Receiver panel assembly estimate data

9.9.1 Assemble and erect receiver panels sheet 1

Description	MH	Unit
Receiver panel assemblies		
Top adjust assembly	8	EA
Bolt connection	0.25	EA
Lower adjust assembly	8	EA
Bolt connection	0.25	EA
Shim plates	2	EA
Linkage attachment	4	EA
Bolt connection	0.25	EA
Linkage assembly	10	EA
Panel	80	EA

9.10 Receiver panel assembly installation estimate

9.10.1 Assemble and erect receiver panels sheet 1

Description	Historical			Estimate		
	MH	Qty	Unit	Qty	Unit	IW
Receiver panel assemblies						**0**
Top adjust assembly	8	14	EA	0	EA	0
Bolt connection	0.25	84	EA	0	EA	0
Lower adjust assembly	8	14	EA	0	EA	0
Bolt connection	0.25	84	EA	0	EA	0
Shim plates	2	14	EA	0	EA	0
Linkage attachment	4	14	EA	0	EA	0
Bolt connection	0.25	56	EA	0	EA	0
Linkage assembly	10	14	EA	0	EA	0
Panel	80	14	EA	0	EA	0

9.11 General scope of field work required for thermal enclosure

Thermal enclosure
External frame assemblies
Back blanket assemblies
Side blanket
Gap blanket
Vert spring hanger boots
Spring hanger
Boot header pipe
Bolt connection

9.12 Receiver thermal enclosure estimate data

9.12.1 Assemble and erect thermal enclosure sheet 1

Description	MH	Unit
Thermal enclosure		
External frame assemblies	17.2	EA
Back blanket assemblies	4.0	EA
Side blanket	4.0	EA
Gap blanket	4.0	EA
Vert spring hanger boots	2.0	EA
Spring hanger	2.0	EA
Boot header pipe	4.0	EA
Bolt connection	0.1	EA

9.13 Receiver thermal enclosure installation estimate

9.13.1 Assemble and erect thermal enclosure sheet 1

Description	MH	Historical Qty	Unit	Estimate Qty	Unit	IW
Thermal enclosure						**0**
External frame assemblies	4.4	30	EA	0	EA	0
Back blanket assemblies	4.0	84	EA	0	EA	0
Side blanket	4.0	28	EA	0	EA	0
Gap blanket	4.0	28	EA	0	EA	0
Vert spring hanger boots	2.0	56	EA	0	EA	0
Spring hanger	2.0	28	EA	0	EA	0
Boot header pipe	4.0	28	EA	0	EA	0
Bolt connection	0.1	13875	EA	0	EA	0

9.14 General scope of field work required for heat shield set

Heat shield sets
Structural—connect to heat shield
Upper heat shield sets
Structural—connect to heat shield
Lower heat shield sets
Corrugated steel panel
Door
Bolt connection
Joint between heat shields
Upper heat shield support rings
W16x31, W6x12, C15x33.9
W16x31, W6x12, C15x33.9
W16x31, W6x12, C15x33.9
W16x31, W6x12, C15x33.9
Field seams
Bolt connection

9.15   Receiver heat shields set estimate data

9.15.1   Assemble and erect heat shields set sheet 1

Description	MH	Unit
Heat shields sets		
Structural—connect to heat shield	2	EA
Upper heat shield sets	40	EA
Structural—connect to heat shield	4	EA
Lower heat shield sets	40	EA
Corrugated steel panel	4	EA
Door	10	EA
Bolt connection	1	EA
Joint between heat shields		
Upper heat shield support rings	4	EA
W16x31, W6x12, C15x33.9	4	EA
W16x31, W6x12, C15x33.9	4	EA
W16x31, W6x12, C15x33.9	4	EA
W16x31, W6x12, C15x33.9	4	EA
Field seams	2	EA
Bolt connection	1	EA

9.16   Receiver heat shields set installation estimate

9.16.1   Assemble and erect heat shields set sheet 1

Description	MH	Qty	Unit	Estimate Qty	Unit	IW
Heat shield sets						0
Structural—connect to heat shield	2	336	EA	0	EA	0
Upper heat shield sets	40	14	EA	0	EA	0
Structural—connect to heat shield	4	24	EA	0	EA	0
Lower heat shield sets	40	14	EA	0	EA	0
Corrugated steel panel	4	14	EA	0	EA	0
Door	10	1	EA	0	EA	0
Bolt connection	1	756	EA	0	EA	0
Joint between heat shields						0
Upper heat shield support rings	4	28	EA	0	EA	0
W16x31, W6x12, C15x33.9	4	30	EA	0	EA	0
W16x31, W6x12, C15x33.9	4	30	EA	0	EA	0
W16x31, W6x12, C15x33.9	4	30	EA	0	EA	0
W16x31, W6x12, C15x33.9	4	30	EA	0	EA	0
Field seams	2	140	EA	0	EA	0
Bolt connection	1	240	EA	0	EA	0

9.17 General scope of field work required for cranes, vents and trays

Set up crane
Install and seal weld exhaust vent curbs
Install exhaust vent
Thermal enclosure trays

9.18 Receiver cranes, vents and trays estimate data

9.18.1 Assemble and erect cranes, vents and trays sheet 1

Description	MH	Unit
Set up crane	240	Job
Install and seal weld exhaust vent curbs	35	EA
Install exhaust vent	20	EA
Thermal enclosure trays	4	EA

9.19 Receiver cranes, vents and trays installation estimate

9.19.1 Assemble and erect cranes, vents and trays sheet 1

		Historical			Estimate		
Description	MH	Qty	Unit	Qty	Unit	IW
Set up crane	240	1	Job	0	Job	0
Install and seal weld exhaust vent curbs	35	4	EA	0	EA	0
Install exhaust vent	20	4	EA	0	EA	0
Thermal enclosure trays	4	240	EA	0	EA	0

9.20 Receiver panel assemblies installation man hours

	Estimated	Actual
	IW	IW
Receiver panel assemblies	1624	0
Thermal enclosure	2360	0
Heat shields sets	2710	0
Joint between heat shields	1112	0
Set up crane	240	0
Install and seal weld exhaust vent curbs	140	0
Install exhaust vent	80	0
Thermal enclosure trays	960	0
Receiver panel assemblies installation man-hours	**9226**	**0**

9.21 General scope of field work required for each air cooled condenser

Air Cooled Condenser
Scope of Work-Field Erection
Off-load and store ACC equipment (transport material to work area)
Erect structural steel incl. stairs and access platforms
Erect structural steel modules 1–6
Fan duct support steel modules 1–6
Preassemble stairways at grade and set and secure
Preassemble external walkways at grade and set and bolt up
Upper structure A-frame steelwork
Field assemble fan deck incl. fans, gearbox's and fan cowls
Fan inlet cowl and fan screens
Motor support bridge
Erect and install wind wall
Erect and weld tube bundles
Erect and weld condensate headers
Field assemble street steam duct
Erect and weld interconnecting street pipe
Erect weld interconnecting equipment pipe
Field assemble transition duct to steam turbine
Field assemble street steam duct
Field assemble risers
Erect and weld out risers
Set and assemble shop tanks
Set and assemble pumps
Erect and weld out transition to steam turbine
Set and connect steam jet air ejectors
Pneumatic testing

9.22 Air cooled condenser estimate data

9.22.1 Structural steel module/fan duct, stairways & walkways, A-frame & fan cowl/screens sheet 1

Description	MH	Unit
Erect and temporarily bolt structural steel	2.50	TON
Bolt connection	0.15	EA
Moment weld	4.00	EA
Handle and place transverse	5.00	EA
Baseplate	2.00	EA
Grout baseplate	1.40	SF
Preassemble stairways at grade and set and secure		
Unload, assemble, and install frame	0.10	SF
Install handrail	0.20	LF
Install stairs	0.85	LF
Install grading	0.17	SF
Preassemble external walkways at grade, set, and bolt up		
Preassemble walkways, set, and bolt up	0.20	SF
Upper structure A-frame steelwork		
Seal plate	2.00	EA
Access door frame	10.00	EA
Cut sheeting to size	0.002	SF
Place sheeting	0.23	SF
Fix sheeting with flashing and caulk	0.05	SF
Tack or screw sheeting	1.00	EA
Install access door	10.00	EA
Pipe hanger with turnbuckle and clevis	10.00	EA
Install steam saddles	10.00	EA
Bolt connection	0.15	EA
Fan inlet cowl and fan screens		
Fan screen supports	2.0	EA
Fan screen modules	3.0	EA
Mount fan cowl segments, stiffening plates, and top angle ring	2.0	EA
Install fan deck plates	2.0	EA

9.22.2 Motor bridge, tube bundles, condensate headers & steam distribution manifolds sheet 2

Description	MH	Unit
Motor support bridge		
20″ plate	2.2	EA
Place fan motor and gearbox	4.0	EA
Bolt fan hub	4.0	EA
Install fan blades	3.0	EA
Lift assembly onto fan deck	4.0	EA
Bolt bridge to structure	4.0	EA
Erect tube bundles		
Remove outlet/inlet tube sheet plate	4.0	EA
Unbolt adjusting plates and install and air seals	4.0	EA
Fix bundle keeper and secondary bundle air seals	4.0	EA
Lift and set tube bundles and tack to CCM bundles	12.0	EA
Weld tube bundles		
Weld inlet tube sheets together (.04″ BW)	0.60	LF
Weld condensate manifold to outlet tube plates	0.75	LF
Weld CCM to lower tube sheets (.79″ BW)	1.20	LF
Erect condensate headers		
Install condensate headers	10.0	EA
Install saddles	2.0	EA
Bolt connection to A-frame	2.3	EA
Weld condensate headers		
16″ schedule 10 BW CS	12.0	JT
18″ schedule 10 BW CS	13.5	JT
Erection of steam distribution manifolds		
Handle/set duct	5.6	EA
Install platform	6.0	EA
Install fixed and sliding saddles	1.0	EA
Make up stiffener rings	2.0	EA
Movable ladder	2.0	EA
Install end plate	4.0	EA
Weld steam distribution manifolds		
6.5′–8.5′ diameter × 5/16″ FW	0.30	LF

9.22.3 IC equipment & street pipe, wind walls, turbine exhaust duct sheet 3

Description	MH	Unit
Erect and weld IC equipment and street pipe		
Handle pipe	0.050	DI
Std BW	0.50	DI
Instrument	2.00	EA
Valves	0.65	DI
Pipe support	1.0	DI
Hydro	0.11	LF
Erection and sheeting of wind walls		
Place sheeting	0.23	SF
Fix sheeting with flashing and caulk	0.05	SF
Tack or screw sheeting	1.00	EA
Erect and weld turbine exhaust duct		
Handle 48″ diameter duct	10.00	EA
Weld duct; 48″ diameter × .75″ WT BW	1.15	LF
48″ Diameter manhole	1.7	DI
Weld in nozzle with reinforcement pad	2.25	LF
Install and align embedded plates	8.0	EA
Weld together saddles and weld stiffeners	4.0	EA
Install saddles	4.0	EA
Bolt saddles to foundation	2.0	EA
Place duct sections	10.0	EA
Grout support saddles	8.0	EA
Weld vanes into elbow		
Install partition plate	80.0	EA
Weld partition plate .86″ BW	1.15	LF
Install vanes	4	EA
Weld vanes fillet weld	2	EA
Cut lifting lug	2	EA
Instrumentation	4	EA
Remove blank plate	0.15	EA
Install stiffener plate	0.6	EA
10″ s80 BW	7.5	EA
10″ s80 fillet weld	7.5	EA

9.22.4   Street steam duct sheet 4

Description	MH	Unit
Field assemble street steam duct		
Handle 20′–6″ diameter duct	10.00	EA
Handle drain pot	10.00	EA
Erect platform	10.00	EA
Field weld with slip ring .75″ BW	1.15	LF
Weld duct; 20′–6″ diameter × .75″ WT BW	1.15	LF
Fit up and Weld .75″ BW	1.15	LF
Install and align embedded plates	8.00	EA
Weld together saddles and weld stiffeners	4.00	EA
Install saddles	4.00	EA
Bolt saddles to foundation	2.00	EA
Place duct sections	10.00	EA
TED-XX handle duct	10.00	EA
Grout support saddles	8.00	EA
Reinforcement pad	1.15	LF

9.22.5   Erect & weld turbine exhaust duct risers sheet 5

Description	MH	Unit
Erection and welding of turbine exhaust duct risers		
Erect duct section, duct with elbow	10.00	EA
Handle duct section, duct section with elbow	10.00	EA
Fit up and weld .75″ BW		
FFW	1.15	LF
FW	1.15	LF
Install and align embedded plates	8.00	EA
Weld together saddles and weld stiffeners	4.00	EA
Install saddles	4.00	EA
Bolt saddles to foundation	2.00	EA

9.22.6   Risers, shop tanks sheet 6

Description	MH	Unit
Field assemble risers		
Handle duct	10.0	EA
Expansion joint 8′–6″	10.0	EA
Set and assemble shop tanks		
Set condensate tank	40.0	EA
Set deaerator tank		
3′–11″ diameter	20.0	EA
8′–6″ diameter	20.0	EA
Fit up and weld		
.5″ x25	0.75	LF
.5″ x54	0.75	LF
Perforated plate	20.0	EA
Wire mesh	20.0	EA

9.22.7   Pumps, expansion joint, test sheet

Description	MH	Unit
Pumps		
Set pump skid and pumps	120.0	EA
Erection and welding of expansion joint (dog bone) at turbine connection		
Weld together upper dog bone holder (including internal braces)		
Handle holder plates	4.0	EA
Handle corner section	4.0	EA
Fit up and weld holder plates 1.8″ thick	4.0	EA
Fit up and weld corner plates 1.1/16″ thick and 7/8″ thick	4.2	LF
Internal braces 5″ diameter std pipe	4.0	EA
Fillet weld	2.0	EA
Set upper dog bone to steam-turbine duct	40.0	EA
Fit up and weld to SPX connection 3/4″ thick	1.1	LF
Fit up and weld SPX to turbine exhaust duct 2 1/4″ thick	2.5	LF
Install dog bone templates	2.0	EA
Rod bar	2.0	EA
Remove steel dog bone templates	2.0	EA
Install rubber and dog bone	1.0	EA
Clamping piece splice	1.0	EA
Field erection bolts	0.2	EA
Pneumatic testing		
Pneumatic testing	1200	TEST

9.23 Air cooled condenser installation estimate

9.23.1 Structural steel module/fan duct, stairways & walkways, A-frame & fan cowl/screens sheet 1

	Historical			Estimate				
Description	MH	Qty	Unit	Qty	Unit	BM	IW	MW
Erect module structural steel						0	0	0
Erect and temporarily bolt structural steel	2.50	11.26	TON	0.00	TON		0	
Bolt connection	0.15	84.00	EA	0.00	EA		0	
Moment weld	4.00	8.00	EA	0.00	EA		0	
Handle and place transverse	5.00	24.00	EA	0.00	EA		0	
Baseplate	2.00	10.00	EA	0.00	EA		0	
Grout baseplate	1.40	36.00	SF	0.00	SF		0	
Fan duct support steel module						0	0	0
Erect and temporarily bolt structural steel	2.50	1.50	TON	0.00	TON		0	
Bolt connection	0.15	1.33	EA	0.00	EA		0	
Preassemble stairways at grade and set						0	0	0
Unload, assemble, and install frame	0.10	19.94	SF	0.00	SF		0	
Install handrail	0.20	8.39	LF	0.00	LF		0	
Install stairs	0.85	3.78	LF	0.00	LF		0	
Install grading	0.17	15.67	SF	0.00	SF		0	
Erect and temporarily bolt structural steel	2.50	0.42	TON	0.00	TON		0	
Grout baseplate	1.40	0.22	SF	0.00	SF		0	
Bolt connection	0.15	2.53	EA	0.00	EA		0	
Baseplate	2.00	0.06	EA	0.00	EA		0	
Preassemble walkways at grade and set						0	0	0
Preassemble walkways, set, and Bolt	0.20	94.37	SF	0.00	SF		0	
Upper structure A-frame steelwork						0	0	0
Erect and temporarily bolt structural steel	2.50	8.52	TON	0.00	TON		0	
Seal plate	2.00	20.00	EA	0.00	EA		0	
Access door frame	10.00	1.00	EA	0.00	EA		0	
Cut sheeting to size	0.002	626.00	SF	0.00	SF		0	
Place sheeting	0.23	420.00	SF	0.00	SF		0	
Fix sheeting with flashing and caulk	0.05	626.00	SF	0.00	SF		0	
Tack or screw sheeting	1.00	25.00	EA	0.00	EA		0	
Install access door	10.00	1.00	EA	0.00	EA		0	

Continued

Description	Historical			Estimate				
	MH	Qty	Unit	Qty	Unit	BM	IW	MW
Pipe hanger with turnbuckle and clevis	10.00	2.00	EA	0.00	EA		0	
Install steam saddles	10.00	3.00	EA	0.00	EA		0	
Bolt connection	0.15	74.00	EA	0.00	EA		0	
Fan inlet cowl and fan screens						**0**	**0**	**0**
Fan screen supports	2.00	12.00	EA	0.00	EA		0	
Fan screen modules	3.00	5.00	EA	0.00	EA		0	
Mount fan cowl segments and top ring	2.00	12.00	EA	0.00	EA		0	
Bolt connection	0.15	36.00	EA	0.00	EA		0	
Install fan deck plates	2.00	12.00	EA	0.00	EA		0	

9.23.2 Structural steel module/fan duct, stairways & walkways, A-frame & fan cowl/screens sheet 2

Description	Historical MH	Qty	Unit	Estimate Qty	Unit	BM	IW	MW
Motor support bridge						0	0	0
20″ plate	2.21	6.00	EA	0.00	EA		0	
Place fan motor and gearbox	4.00	1.00	EA	0.00	EA		0	
Bolt fan hub	4.00	1.00	EA	0.00	EA		0	
Install fan blades	3.00	5.00	EA	0.00	EA		0	
Lift assembly onto fan deck	4.00	1.00	EA	0.00	EA		0	
Bolt bridge to structure	4.00	1.00	EA	0.00	EA		0	
Bolt connection	0.15	31.00	EA	0.00	EA		0	
Handrail	0.20	8.00	LF	0.00	LF		0	
Stairs	0.85	8.00	EA	0.00	EA		0	
Erect and temporarily bolt structural steel	2.50	4.53	TON	0.00	TON		0	
Grating	0.17	100.00	SF	0.00	SF		0	
Erect and weld tube bundles						0	0	0
Erect tube bundles						0	0	0
Remove outlet/inlet tube sheet plate	4.00	2.00	EA	0.00	EA	0		
Unbolt plates and install and air seals	4.00	2.00	EA	0.00	EA	0		
Bundle keeper and secondary bundle air seals	4.00	2.00	EA	0.00	EA	0		
Lift and set tube bundles	12.00	2.00	EA	0.00	EA	0		
Weld tube bundles						0	0	0
Weld inlet tube sheets together (.04″ BW)	0.60	42.00	LF	0.00	LF	0		
Remove inlet tube plate protection sheets	4.00	2.00	EA	0.00	EA	0		
Weld condensate manifold	0.75	42.00	LF	0.00	LF	0		
Weld CCM to lower tube sheets (.79″ BW)	1.20	42.00	LF	0.00	LF	0		
Erect and weld condensate headers						0	0	0
Erect condensate headers						0	0	0
Install condensate headers	10.00	7.00	EA	0.00	EA	0		
Install saddles	2.00	13.00	EA	0.00	EA	0		
Bolt connection to A-frame	2.30	13.00	EA	0.00	EA	0		

Continued

Description	Historical			Estimate				
	MH	Qty	Unit	Qty	Unit	BM	IW	MW
Weld condensate headers						0	0	0
16″ Schedule 10 BW CS	12.00	2.00	JT	0.00	JT	0		
18″ Schedule 10 BW CS	13.50	4.00	JT	0.00	JT	0		
Erect and weld steam distribution manifolds						0	0	0
Erection of steam distribution manifolds						0	0	0
Handle/set duct	5.60	7.00	EA	0.00	EA			0
Install platform	6.00	1.00	EA	0.00	EA			0
Install fixed and sliding saddles	1.00	7.00	EA	0.00	EA			0
Make up stiffener rings	2.00	7.00	EA	0.00	EA			0
Movable ladder	2.00	6.00	EA	0.00	EA			0
Install end plate	4.00	1.00	EA	0.00	EA			0
Weld steam distribution manifolds						0	0	0
6.5′–8.5′ Diameter × 5/16″ FW	0.30	368.82	LF	0.00	LF			0

9.23.3 Structural steel module/fan duct, stairways & walkways, A-frame & fan cowl/screens sheet 3

Description	MH	Historical Qty	Unit	Estimate Qty	Unit	BM	IW	MW
Erect and weld IC equipment pipe						0	0	0
Handle pipe	0.05	456.7	DI	0.00	DI			0
Standard BW	0.50	49.3	DI	0.00	DI			0
Instrument	2.00	0.4	EA	0.00	EA			0
Valves	0.65	0.8	DI	0.00	DI			0
Pipe support	1.00	21.8	DI	0.00	DI			0
Hydro	0.11	99.1	LF	0.00	LF			0
Erect and weld IC street pipe						0	0	0
Handle pipe	0.05	1288.0	DI	0.00	DI			0
Standard BW	0.50	76.0	DI	0.00	DI			0
Pipe support	1.00	43.3	DI	0.00	DI			0
Hydro	0.11	90.0	LF	0.00	LF			0
Erection and sheeting of wind walls						0	0	0
Erect and temporarily bolt structural steel	2.50	7.9	TON	0.00	TON		0	
Door frame	10.00	3.0	EA	0.00	EA		0	
Bolt connection A-frame	0.15	10.0	EA	0.00	EA		0	
Bolt connection	0.15	154.0	EA	0.00	EA		0	
Place sheeting	0.23	260.0	SF	0.00	SF		0	
Fix sheeting with flashing and caulk	0.05	480.0	SF	0.00	SF		0	
Tack or screw sheeting	1.00	16.0	EA	0.00	EA		0	
Erect and weld turbine exhaust duct						0	0	0
Handle 48″ diameter duct	10.00	0.9	EA	0.00	EA	0		
Weld duct; 48″ diameter × .75″ WT BW	1.15	5.9	LF	0.00	LF	0		
48″ diameter manhole	1.70	1.4	DI	0.00	DI	0		
Weld in nozzle with reinforcement pad	2.25	4.9	LF	0.00	LF	0		
Install and align embedded plates	8.00	0.1	EA	0.00	EA	0		
Weld together saddles and weld stiffeners	4.00	1.3	EA	0.00	EA	0		
Install saddles	4.00	0.2	EA	0.00	EA	0		
Bolt saddles to foundation	2.00	0.2	EA	0.00	EA	0		
Place duct sections	10.00	0.2	EA	0.00	EA	0		

Continued

Description	Historical			Estimate				
	MH	Qty	Unit	Qty	Unit	BM	IW	MW
Fit up and weld duct sections .75″ BW	1.15	32.0	LF	0.00	LF	0		
Grout support saddles	8.00	0.2	EA	0.00	EA	0		
Handle 48″ diameter duct	10.00	1.4	EA	0.00	EA	0		
Fit up and weld sections together								
TED 16,17 .75″ BW	1.15	3.0	LF	0.00	LF	0		
Place duct sections	10.00	0.3	EA	0.00	EA	0		
Fit up and weld								
TED 16, 17 .75″ BW	1.15	28.5	LF	0.00	LF	0		
TED 17, 15, 7A, 7B .75″ BW	1.15	11.3	LF	0.00	LF	0		
Install partition plate	4.00	2.2	EA	0.00	EA	0		
Weld partition plate .86″ BW	44.88	0.0	LF	0.00	LF	0		
Install vanes	18.00	0.1	EA	0.00	EA	0		
Weld vanes fillet weld	54.00	0.1	EA	0.00	EA	0		
Cut lifting lug	1.00	0.1	EA	0.00	EA	0		
Instrumentation	14.00	0.1	EA	0.00	EA	0		
Remove blank plate	128.00	0.0	EA	0.00	EA	0		
Install stiffener plate	40.00	0.0	EA	0.00	EA	0		
10″ s 80 BW	2.00	0.2	EA	0.00	EA	0		
10″ s 80 fillet weld	4.00	0.2	EA	0.00	EA	0		

9.23.4   Street steam duct sheet 4

Description	Historical			Estimate		
	MH	Qty	Unit	Qty	Unit	BM
Field assemble street steam duct						0
Handle 20′–6″ diameter duct	10.00	1.3	EA	0.00	EA	0
Weld duct; 20′–6″ diameter × .75″ WT BW	1.15	8.3	LF	0.00	LF	0
Handle drain pot	10.00	0.0	EA	0.00	EA	0
Fit up and Weld .75″ BW	1.15	1.6	LF	0.00	LF	0
Reinforcement pad	1.15	3.2	LF	0.00	LF	0
Install and align embedded plates	8.00	0.1	EA	0.00	EA	0
Weld together saddles and weld stiffeners	4.00	1.3	EA	0.00	EA	0
Install saddles	4.00	0.2	EA	0.00	EA	0
Bolt saddles to foundation	2.00	0.3	EA	0.00	EA	0
Erect platform	10.00	0.7	EA	0.00	EA	0
Place duct sections	10.00	0.3	EA	0.00	EA	0
Fit up and weld .75″ BW	1.15	43.1	LF	0.00	LF	0
Grout support saddles	8.00	0.1	EA	0.00	EA	0
Field weld with slip ring .75″ BW	1.15	3.3	LF	0.00	LF	0
TED-06, 07, 08, 09 handle duct	10.00	1.6	EA	0.00	EA	0
Fit up and weld .75″ BW	1.15	9.6	EA	0.00	EA	0
Install and align embedded plates	8.00	0.1	EA	0.00	EA	0
Weld together saddles and weld stiffeners	4.00	1.3	EA	0.00	EA	0
Install saddles	4.00	0.3	EA	0.00	EA	0
Bolt saddles to foundation	2.00	0.3	EA	0.00	EA	0
Place duct sections	10.00	0.4	EA	0.00	EA	0
Fit up and weld .75″ BW	1.15	40.0	LF	0.00	LF	0
Field weld w/slip ring .75″ BW	1.15	5.9	LF	0.00	LF	0
Grout support saddles	8.00	0.1	EA	0.00	EA	0
Place duct sections TED 05, 04	10.00	0.8	EA	0.00	EA	0
Fit up and weld .75″ BW	1.15	4.8	LF	0.00	LF	0
Install and align embedded plates	8.00	0.1	EA	0.00	EA	0
Weld together saddles and weld stiffeners	4.00	0.7	EA	0.00	EA	0
Install saddles	4.00	0.2	EA	0.00	EA	0
Bolt saddles to foundation	2.00	0.2	EA	0.00	EA	0
Place duct	10.00	0.2	EA	0.00	EA	0
Fit up and weld .75″ BW	1.15	18.3	LF	0.00	LF	0
Field weld with slip ring .75″ BW	1.15	2.2	LF	0.00	LF	0
Place duct sections TED 03, 02	10.00	0.8	EA	0.00	EA	0
Fit up and Weld .75″ BW	1.15	4.8	LF	0.00	LF	0
Install and align embedded plates	8.00	0.1	EA	0.00	EA	0
Weld together saddles and weld stiffeners	4.00	0.7	EA	0.00	EA	0
Install saddles	4.00	0.2	EA	0.00	EA	0
Bolt saddles to foundation	2.00	0.2	EA	0.00	EA	0
Place duct	10.00	0.2	EA	0.00	EA	0
Fit up and weld .75″ BW	1.15	13.0	LF	0.00	LF	0
Field weld with slip ring .75″ BW	1.15	1.5	LF	0.00	LF	0
Grout support saddles	8.00	0.1	EA	0.00	EA	0

9.23.5   Erect & weld turbine exhaust duct risers sheet 5

Description	Historical			Estimate				
	MH	Qty	Unit	Qty	Unit	BM	IW	MW
Erect and weld turbine exhaust duct risers								
Erect and weld out risers						0	0	0
Erect duct section	10.00	0.14	EA	0.00	EA	0		
Fit up and weld .75″ BW								
FFW	1.15	7.50	LF	0.00	LF	0		
FW	1.15	7.50	LF	0.00	LF	0		
Erect duct with elbow	10.00	0.14	EA	0.00	EA	0		
Fit up and weld .75″ BW								
FFW	1.15	7.50	LF	0.00	LF	0		
FW	1.15	7.50	LF	0.00	LF	0		
Handle duct section with elbow	10.00	0.03	EA	0.00	EA	0		
Fit up and weld .75″ BW	1.15	1.50	LF	0.00	LF	0		
Install and align embedded plates	8.00	0.11	EA	0.00	EA	0		
Weld together saddles and weld stiffeners	4.00	1.33	EA	0.00	EA	0		
Install saddles	4.00	0.33	EA	0.00	EA	0		
Bolt saddles to foundation	2.00	0.11	EA	0.00	EA	0		
Handle duct section	10.00	0.08	EA	0.00	EA	0		
Handle duct section with elbow	10.00	0.03	EA	0.00	EA	0		
Fit up and weld .75″ BW								
FFW	1.15	1.50	LF	0.00	LF	0		
FW	1.15	1.50	LF	0.00	LF	0		

9.23.6   Risers, tanks sheet 6

Description	Historical			Estimate				
	MH	Qty	Unit	Qty	Unit	BM	IW	MW
Field assemble risers						0	0	0
Duct diameter 8'–6"								
Handle TED duct	10.00	0.14	EA	0.00	EA	0		
Fit up and weld .75" BW	1.15	6.94	LF	0.00	LF	0		
Handle 7051RD2	10.00	0.14	EA	0.00	EA	0		
Handle 7052RD2	10.00	0.14	EA	0.00	EA	0		
Fit up and weld .75" BW	1.15	7.50	LF	0.00	LF	0		
Expansion joint 8'–6"	10.00	0.28	EA	0.00	EA	0		
Handle 7021RD1	10.00	0.14	EA	0.00	EA	0		
Handle 7011TE2 and 7011EV2 6'–10"	10.00	0.28	EA	0.00	EA	0		
Handle 7011TE1 and 7011EV1 6'–10"	10.00	0.28	EA	0.00	EA	0		
Fit up and weld .75" BW	1.15	7.50	LF	0.00	LF	0		
Handle TED 1	10.00	0.03	EA	0.00	EA	0		
Expansion joint 8'–6"	10.00	0.03	EA	0.00	EA	0		
Fit up and weld .75" BW	1.15	3.00	LF	0.00	LF	0		
Handle TED 1 6'–10"	10.00	0.03	EA	0.00	EA	0		
Handle 7046RD1	10.00	0.03	EA	0.00	EA	0		
Handle 7026RD2	10.00	0.03	EA	0.00	EA	0		
Fit up and weld .75" BW	1.15	1.50	LF	0.00	LF	0		
Expansion joint 8'–6"	10.00	0.06	EA	0.00	EA	0		
7026RD1 handle 6–10"	10.00	0.03	EA	0.00	EA	0		
7016TE2 and 7016TE 1 6'–10"	10.00	0.06	EA	0.00	EA	0		
7016EV1 and 7016EV2 6'–10"	10.00	0.06	EA	0.00	EA	0		
Fit up and weld .75" BW	1.15	7.50	LF	0.00	LF	0		
Set and assemble shop tanks						0	0	0
Install and align embedded plates	8.00	0.11	EA	0.00	EA	0		
Weld together saddles and weld stiffeners	4.00	0.33	EA	0.00	EA	0		
Install saddles	4.00	0.06	EA	0.00	EA	0		
Bolt saddles to foundation	2.00	0.06	EA	0.00	EA	0		
Baseplates	2.00	0.06	EA	0.00	EA	0		
Set condensate tank	40.00	0.03	EA	0.00	EA	0		
Set deaerator tank								
3'–11" diameter	20.00	0.03	EA	0.00	EA	0		
8'–6" diameter	20.00	0.03	EA	0.00	EA	0		
Fit up and weld								
.5" × 25	0.75	0.69	LF	0.00	LF	0		
.5" × 54	0.75	1.50	LF	0.00	LF	0		
Perforated plate	20.00	0.06	EA	0.00	EA	0		
Wire mesh	20.00	0.06	EA	0.00	EA	0		

9.23.7   Pumps, expansion joint, test sheet 7

Description	Historical			Estimate				
	MH	Qty	Unit	Qty	Unit	BM	IW	MW
Pumps						0	0	0
Set pump skid and pumps	120.0	1.00	EA	0.00	EA	0		
Install expansion joint at turbine						0	0	0
Weld together upper dog bone holder								
Handle holder plates	4.0	0.56	EA	0.00	EA	0		
Handle corner section	4.0	0.22	EA	0.00	EA	0		
Fit up and weld holder plates 1.8″ thick	4.0	0.61	EA	0.00	EA	0		
Weld corner plates 1.1/16″ thick and 7/8″ thick	4.2	0.49	LF	0.00	LF	0		
Internal braces 5″ diameter std pipe	4.0	0.61	EA	0.00	EA	0		
Fillet weld	2.0	1.33	EA	0.00	EA	0		
Set upper dog bone to steam-turbine duct	40.0	0.06	EA	0.00	EA	0		
Fit up and weld to SPX connection 3/4″ thick	1.1	4.77	LF	0.00	LF	0		
Weld SPX to turbine exhaust duct 2 1/4″ thick	2.5	4.77	LF	0.00	LF	0		
Install dog bone templates	2.0	0.22	EA	0.00	EA	0		
Rod bar	2.0	0.44	EA	0.00	EA	0		
Remove steel dog bone templates	2.0	0.22	EA	0.00	EA	0		
Install rubber at dog bone	1.0	4.77	EA	0.00	EA	0		
Clamping piece splice	1.0	2.96	EA	0.00	EA	0		
Field erection bolts	0.2	11.56	EA	0.00	EA	0		
Pneumatic testing						0	0	0
Pneumatic testing	1200	0.03	TEST	0.00	TEST	0		

9.24   Air cooled condenser installation man hours

9.24.1   Facility—solar power plant (10 cell ACC)

Description	Historical			Estimate	Actual
				MH/	MH/
	BM	IW	PF	Module	Module
Erect structural steel modules 1–6	0.0	0.0	0.0	**0.0**	263.1
Fan duct support steel modules 1–6	0.0	0.0	0.0	**0.0**	4.0
Preassemble stairways at grade and set and secure	0.0	0.0	0.0	**0.0**	11.4
Preassemble external walkways at grade	0.0	0.0	0.0	**0.0**	18.6
Upper structure A-frame steelwork	0.0	0.0	0.0	**0.0**	294.7
Fan inlet cowl and fan screens	0.0	0.0	0.0	**0.0**	92.4
Motor support bridge	0.0	0.0	0.0	**0.0**	85.6
Erect and weld tube bundles	0.0	0.0	0.0	**0.0**	163.1
Erect and weld condensate headers	0.0	0.0	0.0	**0.0**	203.9
Erect and weld steam distribution manifolds	0.0	0.0	0.0	**0.0**	192.8
Erect and weld IC equipment pipe	0.0	0.0	0.0	**0.0**	81.7
Erect and weld IC street pipe	0.0	0.0	0.0	**0.0**	155.7
Erection and sheeting of wind walls	0.0	0.0	0.0	**0.0**	173.0
Erect and weld turbine exhaust duct	0.0	0.0	0.0	**0.0**	163.3
Field assemble street steam duct	0.0	0.0	0.0	**0.0**	272.4
Erect and weld out risers	0.0	0.0	0.0	**0.0**	107.9
Set and assemble shop tanks	0.0	0.0	0.0	**0.0**	8.8
Pumps	0.0	0.0	0.0	**0.0**	120.0
Install expansion joint at turbine	0.0	0.0	0.0	**0.0**	43.3
Pneumatic testing	0.0	0.0	0.0	**0.0**	33.3
Air cooled condenser installation man-hours	**0**	**0**	**0**	**0.0**	**2489.0**
Actual number of streets × modules = number of cells	**2**	**5**	**10**		
MH/module × number of cells = estimate man-hours	**2489**	**10**	**24,890**		
Estimate number of streets × modules = number of cells	**0**	**0**	**0**		
MH/module × number of cells = estimate man-hours	**0**	**0**	**0**		

9.25 Solar plant equipment man hour breakdown

	Actual	Estimate			
Facility—solar plant equipment	MH	BM	PF	IW	MH
Receiver installation man-hours	12,768			0	0
Receiver panel assemblies installation man-hours	9226			0	0
Air cooled condenser (10 cell)—installation man-hours	24,890	0	0	0	0
Solar plant project man-hours	**46,884**	**0**	**0**	**0**	**0**

Chapter 10

Boiler tube replacement

10.1 General scope of field work required for boiler tube replacement sheet 1

Scope of Work-Field Erection
Boiler scaffolding and rigging
Open steam drum and/or lower drum
Remove drum internals
Remove soot blowers
Remove boiler trim
Remove casing, tie bars, and brill
Remove existing tubes (cut and remove)
Knock out tube butts and clean tube holes
Install new tubes
Expand tubes in steam drum and lower drum
Reinstall boiler casing (including tie bars)
Reinstall boiler trim
Reinstall soot blowers
Close steam and lower drums
Testing
Clean up and remove boiler scaffold and rigging
Disconnect and reconnect gas outlet

Industrial Process Plant Construction Estimating and Man-Hour Analysis.
https://doi.org/10.1016/B978-0-12-818648-0.00010-7

10.2 Boiler tube replacement estimate data

10.2.1 Boiler tube replacement sheet 1

Description	MH	Unit
Boiler scaffolding and rigging	64.0	Boiler
Open steam drum and/or lower drum	4.0	Man way
Remove drum internals	64.0	Boiler
Remove sootblowers	20.0	EA
Remove boiler trim	40.0	Boiler
Remove casing, tie bars, and brill	0.5	SF
Remove existing tubes (cut and remove)	1.0	Tube
Knock out tube butts and clean tube holes	0.5	Butt
Install new tubes	1.0	Tube
Expand tubes 2″ OD; in steam drum and lower drum	0.36	EA
Reinstall boiler casing (including tie bars)	1.5	LF
Reinstall boiler trim	48.0	Boiler
Reinstall soot blowers	32.0	EA
Close steam and lower drums	4.0	Man way
Testing	56.0	Boiler
Clean up and remove boiler scaffold and rigging	80.0	Boiler
Disconnect and reconnect gas outlet	40.0	Boiler

10.3 Boiler tube replacement installation estimate

10.3.1 Boiler tube replacement sheet 1

	Historical		Estimate		
	MH	Unit	Qty	Unit	BM
Boiler tube replacement					0
Boiler scaffolding and rigging	64.0	Boiler	0	Boiler	0
Open steam drum and/or lower drum	4.0	Man way	0	Man way	0
Remove drum internals	64.0	Boiler	0	Boiler	0
Remove sootblowers	20.0	EA	0	EA	0
Remove boiler trim	40.0	Boiler	0	Boiler	0
Remove casing, tie bars, and brill	0.5	SF	0	SF	0
Remove existing tubes (cut and remove)	1.0	Tube	0	Tube	0
Knock out tube butts and clean tube holes	0.5	Butt	0	Butt	0
Install new tubes	1.0	Tube	0	Tube	0
Expand tubes 2″ OD; in steam drum and lower drum	0.36	EA	0	EA	0
Reinstall boiler casing (including tie bars)	1.5	LF	0	LF	0
Reinstall boiler trim	48.0	Boiler	0	Boiler	0
Reinstall sootblowers	32.0	EA	0	EA	0
Close steam and lower drums	4.0	Man way	0	Man way	0
Testing	56.0	Boiler	0	Boiler	0
Clean up and remove boiler scaffold and rigging	80.0	Boiler	0	Boiler	0
Disconnect and reconnect gas outlet	40.0	Boiler	0	Boiler	0

10.4   Boiler tube replacement installation man hours

	Estimated	Actual
	BM	BM
Boiler tube replacement	0	0
Boiler tube replacement installation man hours	**0**	**0**

Chapter 11

Sample estimates

11.1 Section introduction

This section provides the **industrial process plant construction estimating process** to enable the reader to use the statistical and estimating methods, scopes of work, man-hour tables, and estimate sheets to

(1) evaluate the accuracy and verify historical data collected in the field for process piping and equipment installed in industrial process plants;

(2) provide a comprehensive and accurate method, using construction statistics, and estimating methods to compile detailed estimates, RFPs, and field change orders.

The section enables the reader to use sample estimating forms for piping and equipment to set up detailed estimates on a desktop computer.

Equipment estimates illustrate the steps required to calculate the detailed craft man-hour estimate. Examples will illustrate how the unit quantity model, data, and tables that are summed in the computer model the erection sequence.

The sample estimate does not include cost and man-hours for material, equipment usage, indirect craft and supervision, project staff, warehousing and storage, shop fabrication, overheads, and fee. The direct craft man-hour estimate is the basis for the estimator to obtain the project schedule and the man-hours and cost for indirect craft and supervision, project staff, construction equipment, material, subcontractors, mobilization and demobilization, site general conditions, overhead, and fee. In addition, the estimator must determine all factors that will affect direct craft labor productivity and overtime impacts.

The estimator can use the table of labor factors and values for factoring labor productivity in the introduction.

Industrial Process Plant Construction Estimating and Man-Hour Analysis.
https://doi.org/10.1016/B978-0-12-818648-0.00011-9
221

11.2 Industrial process plant construction estimating process

Process plant labor estimating is a process that includes the four steps:

(1) The first step starts with the quantity takeoff, field-specific scope.
(2) Step two converts the field-specific scope to quantities using the comparison method; takeoff represents the complete scope of work.
(3) The third step is the application of man-hour units to the quantified scope of work. The man-hour data tables have been developed from historical data, collected from field cost reports, and verified using graphic and regression analysis.
(4) Final step is to enter the unit man-hour rates from the tables and the quantities from the "takeoff" into the estimate sheet and calculate the direct craft man-hours using the unit quantity method (man-hour rate × "takeoff" quantities). The estimate man-hours are compared with the historical man-hours.

11.3 Sample estimate construction estimating process

Table 11.3.1 Example field erection fan inlet cowl and fan screens

(1) The first step starts with the quantity takeoff, field-specific scope.
Fan inlet cowl and fan screens
Fan screen supports
Fan screen modules
Mount fan cowl segments and top ring
Bolt connection
Install fan deck plates
(2) Convert the field-specific scope to quantities using the comparison method.
(3) The third step is the application of man-hour units to the quantified scope of work. The man-hour data tables have been developed from historical data, collected from field cost reports, and verified using graphic analysis.

TABLE 11.3.1 MH_a = estimate for increased fan inlet cowl and fan screens quantities

Estimate—ACC-solar power plant-fan inlet cowl screen	Historical		Estimate	
Description	*Qty*	*Unit*	*Qty*	*Unit*
Fan inlet cowl and fan screens				
Fan screen supports	12.00	EA	15.00	EA
Fan screen modules	5.00	EA	6.25	EA
Mount fan cowl segments and top ring	12.00	EA	15.00	EA
Bolt connection	36.00	EA	45.00	EA
Install fan deck plates	12.00	EA	15.00	EA

Project: Air-cooled condenser—fan inlet cowl and fan screens
Foreman: John Smith   Date: 0000 June 00
Craft: Boilermaker

Cost code	Phase code description	MH	Qty
	Fan inlet cowl and fan screens		
000000	Fan screen supports	2.00	12.00
000000	Fan screen modules	3.00	5.00
000000	Mount fan cowl segments and top ring	2.00	12.00
000000	Bolt connection	0.15	36.00
000000	Install fan deck plates	2.00	12.00

Graphic analysis of data

Use Excel's chart capabilities to plot the graph for the simple moving average forecast.

To use the Excel chart capabilities, highlight the range E42-D52, and select **Insert, Scatter Chart, Chart Elements, Axis, Axis Titles, Chart Title, Gridlines, Legend, Trendline, and More Options: Display equation on chart and Display R-squared value on chart.**

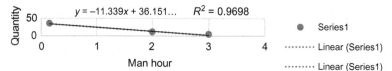

FIG. 11.3.1   Fan inlet cow and fan screens.

The coefficient of determination is $R^2 = 0.9698$, and the correlation coefficient, $R = 0.9847$, is a strong indication of correlation (Fig. 11.3.1). 98.4% of the total variation on Y can be explained by the linear relationship between X and Y (described by the regression equation, $Y = 11.339x + 36.151$). The relationship between X and Y variables is such.

Man-hour table...fan inlet cowl and fan screens

Description	MH	Unit
Fan inlet cowl and fan screens		
Fan screen supports	2.0	EA
Fan screen modules	3.0	EA
Mount fan cowl segments and top ring	2.0	EA
Bolt connection	0.15	EA
Install fan deck plates	2.0	EA

(4) Final step is to enter the unit man-hour rates from the tables and the quantities from the "takeoff" into the estimate sheet and calculate the direct craft man-hours using the unit quantity method (man-hour rate × "takeoff" quantities). The estimate man-hours are compared with the historical man-hours.

Description	Historical			Estimate					
	MH	Qty	Unit	Qty	Unit	BM	IW	MW
Fan inlet cowl and fan screens						0	115.5	0
Fan screen supports	2.00	12.00	EA	15.00	EA		30.0	
Fan screen modules	3.00	5.00	EA	6.25	EA		18.8	
Mount fan cowl segments and top ring	2.00	12.00	EA	15.00	EA		30.0	
Bolt connection	0.15	36.00	EA	45.00	EA		6.8	
Install fan deck plates	2.00	12.00	EA	15.00	EA		30.0	

Air-cooled condenser installation man-hours					
Facility—solar power plant (10-cell ACC)	Historical			Estimate	Actual
Description	BM	IW	PF	MH/module	MH/module
Fan inlet cowl and fan screens	0.0	115.5	0.0	**115.5**	92.4

11.4   Sample estimate for assembly receiver tower IC pipe

Process piping estimate form

Facility:

Work scope:

Description	Pipe Size	Unit	Qty	Unit MH	Factor	Total MH

Column totals

11.4.1   Handle and install pipe -welded joint sheet 1

Facility: solar plant

Work scope: handle and install
pipe — welded joint

Description	Pipe Size	Unit	Qty	MH/LF	Factor	Total MH
Assembly receiver tower IC pipe						
Cold supply sch 20S SA-312 TP 347H	16	LF	32	1.12	1.00	36
Jumper sch 20S SA-312 TP 347H	16	LF	97	1.12	1.00	109
Crossover sch 20S SA-312 TP 347H	16	LF	83	1.12	1.00	93
Discharge sch 20S SA-312 TP 347H	16	LF	23	1.12	1.00	26
Downcomer sch 20S SA-312 TP 347H	20	LF	88	1.40	1.00	123
Drain sch 10S SA-312 TP 347H	6	LF	48	0.42	1.00	20
Drain header sch 10S SA-312 TP 347H	10	LF	90	0.70	1.00	63
Drain header sch 10S SA-312 TP 347H	6	LF	5	0.42	1.00	2
Receiver vent sch 10S SA-312 TP 347H	10	LF	136	0.70	1.00	95
Circuit 2 vent sch 10S SA-312 TP 347H	4	LF	139	0.28	1.00	39
ECV vent sch 10S SA-312 TP 347H	8	LF	27	0.56	1.00	15
Column totals			736			585

11.4.2   Welding BW, SW, & PWHT—arc uphill sheet 2

Facility: solar plant

Work scope: welding BW, SW,
PWHT—arc uphill

Description	Pipe Size	Unit	Qty	MH/JT	Factor	Total MH
Cold supply sch 20S SA-312 TP 347H	16	EA	4	9.6	2.23	86
Jumper sch 20S SA-312 TP 347H	16	EA	33	9.6	2.23	706
Crossover sch 20S SA-312 TP 347H	16	EA	12	9.6	2.23	257
Discharge sch 20S SA-312 TP 347H	16	EA	4	9.6	2.23	86
Downcomer sch 20S SA-312 TP 347H	20	EA	6	12	2.45	176
Drain sch 10S SA-312 TP 347H	6	EA	24	3.6	1.65	143
Drain header sch 10S SA-312 TP 347H	10	EA	9	6	1.85	100
Drain header sch 10S SA-312 TP 347H	6	EA	5	3.6	1.65	30
Receiver assembly vent sch 10S SA-312 TP 347H	10	EA	6	6	1.85	67
Circuit 2 vent sch 10S SA-312 TP 347H	4	EA	25	2.4	1.61	97
ECV vent sch 10S SA-312 TP 347H	8	EA	6	4.8	1.74	50
Column totals			134			1796

11.4.3   Bolt up of flanged joint by weight class sheet 3

Facility: solar plant

Work scope: bolt up of flanged
joint by weight class

Description	Pipe Size	Unit	Qty	MH/JT	Factor	Total MH
Cold supply sch 20S SA-312 TP 347H	16	EA	0	0	1.00	0
Jumper sch 20S SA-312 TP 347H	16	EA	0	0	1.00	0
Crossover sch 20S SA-312 TP 347H	16	EA	0	0	1.00	0
Discharge sch 20S SA-312 TP 347H	16	EA	1	6.4	1.00	6
Downcomer sch 20S SA-312 TP 347H	20	EA	4	8.0	1.00	32
Drain sch 10S SA-312 TP 347H	6	EA	0	0	1.00	0
Drain header sch 10S SA-312 TP 347H	10	EA	0	0	1.00	0
Drain header sch 10S SA-312 TP 347H	6	EA	0	0	1.00	0
Receiver assembly vent sch 10S SA-312 TP 347H	10	EA	0	0	1.00	0
Circuit 2 vent sch 10S SA-312 TP 347H	4	EA	5	1.6	1.00	8
ECV vent sch 10S SA-312 TP 347H	8	EA	1	3.2	0	0
Column totals			11			46.4

11.4.4   Field handle valves/specialty items by weight class

Facility: Solar plant

Work scope: field handle valves/
specialty items by weight class

Description	Pipe Size	Unit	Qty	MH/EA	Factor	Total MH
Cold supply sch 20S SA-312 TP 347H	16	EA	0	0.00	1.00	0
Jumper sch 20S SA-312 TP 347H	16	EA	0	0.00	1.00	0
Crossover sch 20S SA-312 TP 347H	16	EA	0	0.00	1.00	0
Discharge sch 20S SA-312 TP 347H	16	EA	1	7.20	1.00	7
Downcomer sch 20S SA-312 TP 347H	20	EA	2	9.00	1.00	18
Drain sch 10S SA-312 TP 347H	6	EA	6	2.70	1.00	16
Drain header sch 10S SA-312 TP 347H	10	EA	0	0.00	1.00	0
Drain header sch 10S SA-312 TP 347H	6	EA	0	0.00	1.00	0
Receiver assembly vent sch 10S SA-312 TP 347H	10	EA	1	4.50	1.00	5
Circuit 2 vent sch 10S SA-312 TP 347H	4	EA	2	1.60	1.00	3
ECV vent sch 10S SA-312 TP 347H	8	EA	0	0.00	1.00	0
Column totals			12			49.1

11.4.5   Pipe supports sheet 5

Facility: solar plant

Work scope: pipe supports

Description	Pipe Size	Unit	Qty	MH/EA	Factor	Total MH
Cold supply sch 20S SA-312 TP 347H	16	EA	0	0.00	1.00	0
Jumper sch 20S SA-312 TP 347H	16	EA	0	0.00	1.00	0
Crossover sch 20S SA-312 TP 347H	16	EA	6	9.60	1.00	58
Discharge sch 20S SA-312 TP 347H	16	EA	2	9.60	1.00	19
Downcomer sch 20S SA-312 TP 347H	20	EA	2	12.00	1.00	24
Drain sch 10S SA-312 TP 347H	6	EA	0	0.00	1.00	0
Drain header sch 10S SA-312 TP 347H	10	EA	0	0.00	1.00	0
Drain header sch 10S SA-312 TP 347H	6	EA	0	0.00	1.00	0
Receiver assembly vent sch 10S SA-312 TP 347H	10	EA	7	6.00	1.00	42
Circuit 2 vent sch 10S SA-312 TP 347H	4	EA	7	2.40	1.00	17
ECV vent sch 10S SA-312 TP 347H	8	EA	1	4.80	1.00	5
Column totals			25			164

11.5   Man hours for assembly receiver tower interconnecting pipe

Facility—Solar plant

Equipment—Solar plant

Estimate sheet 7—assembly receiver IC pipe		
Description	PF/MH	MH/LF
Estimate sheet 1—handle and install pipe-welded joint	585	
Estimate sheet 2—welding: BW, SW, PWHT arc uphill	1796	
Estimate sheet 3—bolt up of flanged joint by weight class	46	
Estimate Sheet 4—handle valves by weight class	49	
Estimate Sheet 5—pipe supports	164	
Estimate Sheet 6—instrument	0	
Hydro	211	
Column totals	2853	3.88

11.6   Sample equipment estimate estimates

Sample equipment estimate form
Equipment estimate—Facility

Description	Historical			Estimate				
	MH	Qty	Unit	Qty	Unit	BM	IW	PF
	0.00	0.0		0.0		0	0	0
	0.00	0.0		0.0		0	0	0
	0.00	0.0		0.0		0	0	0
	0.00	0.0		0.0		0	0	0
	0.00	0.0		0.0		0	0	0
	0.00	0.0		0.0		0	0	0
	0.00	0.0		0.0		0	0	0
	0.00	0.0		0.0		0	0	0
	0.00	0.0		0.0		0	0	0
	0.00	0.0		0.0		0	0	0
	0.00	0.0		0.0		0	0	0

Facility

Equipment installation man-hours	Actual	Estimated
Description	MH	BM
	0	0
	0	0
	0	0
	0	0
	0	0
	0	0
	0	0
	0	0
	0	0
Equipment installation man-hours	0	0

11.7 Waste heat boiler, 500,000 lb/h installation estimate

11.7.1 Drums, generating tubes, headers, side, front & rear panels sheet 1

Description	Historical		Estimate		
	MH	Unit	Qty	Unit	BM
Install steam and mud drums					**1245**
Steam drum and drum saddles	3.8	TON	132.6	TON	504
Mud drum and drum saddles	4.6	TON	75.0	TON	345
Steam drum trim piping	260.0	LOT	1.0	LOT	260
Drum internals—remove and reinstall	136.0	Boiler	1.0	Boiler	136
Generating bank tubes					**4354**
2-1/2″ OD × 0.203″ wall thickness, swage, and roll					
Install generating tubes	0.50	EA	3154.8	EA	1577
2′1/2″ Ends—expand tubes in steam and mud drums	0.44	End	6309.6	End	2776
Ground assembly—headers, furnace, and wall panels					**3230**
Fit and weld water wall panels to headers					
Lower side wall header—lower side wall panels					
Place lower side wall header	4.0	TON	13.7	TON	55
Place lower side wall panels—water wall	2.5	TON	56.5	TON	141
Field tube welding—vver 2-1/2″ TIG	4.2	EA	187.2	EA	786
Upper side wall header—upper side wall panels					
Place upper side wall header	4.0	TON	11.1	TON	44
Place upper side wall panels—water wall	2.5	TON	56.5	TON	141
Field tube welding—over 2-1/2″ TIG	4.2	EA	187.2	EA	786
Place front wall header	4.0	TON	6.1	TON	24
Place front wall panels	2.5	TON	28.2	TON	71
Field tube welding—Over 2-1/2″ TIG	4.2	EA	129.6	EA	544
Place rear wall header	4.0	TON	5.3	TON	21
Place rear wall panels—water wall	2.5	TON	28.2	TON	71
Field tube welding—ver 2-1/2″ TIG	4.2	EA	129.6	EA	544

11.7.2 Erect headers & panels, weld water wall tubes, burner, soot blower, superheater sheet 2

Description	Historical			Estimate		
	MH	Unit	Qty	Unit	BM
Erect headers and wall panels, fit and weld wall panels					**5291**
Scaffolding and rigging	136.00	Boiler	1.0	Boiler	136
Erect right-side wall panels—water wall	2.50	Ton	144.0	Ton	360
Fit and weld filler bar at tube welds	1.00	Space	561.6	Space	562
Fit and weld filler joining membrane panels	0.60	LF	192.0	LF	115
Erect left-side wall panels—water wall	2.50	Ton	144.0	Ton	360
Fit and weld filler bar at tube welds	1.00	Space	561.6	Space	562
Fit and weld filler joining membrane panels	0.60	LF	192.0	LF	115
Erect front wall panels—water wall	2.50	Ton	144.0	Ton	360
Fit and weld filler bar at tube welds	1.00	Space	518.4	Space	518
Fit and weld filler joining membrane panels	0.60	LF	187.2	LF	112
Erect rear wall panels—water wall	2.50	Ton	144.0	Ton	360
Fit and weld filler bar at tube welds	1.00	Space	518.4	Space	518
Fit and weld filler joining membrane panels	0.60	LF	187.2	LF	112
Erect roof panels—water wall	2.50	Ton	108.0	Ton	270
Fit and weld filler bar at tube welds	1.00	Space	705.6	Space	706
Fit and weld filler joining membrane panels	0.60	LF	207.6	LF	125
Fit and weld water wall tubes	4.20	EA	1975.0	EA	**8295**
Erect, fit, and weld primary/secondary headers					**1817**
Primary super heater headers/coils	4.20	Ton	115.9	Ton	487
10.75″ diameter inlet header	4.00	Ton	2.5	Ton	10
20″ outlet header	4.00	Ton	6.0	Ton	24
Weld super heater tubes; 1.75″ OD	3.70	EA	350.4	EA	1296
Burner system					**492**
Burner and wind box	59.40	Ton	5.7	Ton	337
Fan with drive	45.00	Ton	2.1	Ton	96
Silencer	60.00	EA	1.0	EA	60
Sootblowers					**1280**
Sootblower					
Install rotary and retractable sootblowers					
Super heater—sootblower with pipe, valve, fitting	64.00	EA	12.0	EA	768
Economizer—sootblower with pipe, valve, fitting	64.00	EA	4.0	EA	256
Generating bank—sootblower with PVF	64.00	EA	4.0	EA	256

11.7.3   Down comers & code piping sheet 3

Description	Historical MH	Unit	Estimate Qty	Unit	BM
Erect and install downcomers and code piping					**1811**
Handle downcomers, 12″ OD × 0.562 WT × 40′	1.32	LF	336.0	LF	444
Downcomers, 12″ OD × 0.562 WT BW	12.6	EA	16.8	EA	212
Handle 12″ 300# BW main stop valve and ck valve	5.4	EA	2.4	EA	13
12″ × 0.562 WT CS BW	12.60	EA	4.8	EA	60
Handle 10″ 300# BW feed water stop and ck valve	136.00	EA	2.4	EA	326
10″ × 0.562 WT CS BW	4.00	EA	4.8	EA	19
Handle 8″ 300# BW spray control valves	7.20	EA	4.8	EA	35
8″ × 0.562 WT CS BW	8.40	EA	9.6	EA	81
Handle 3″ 300# safety valves	2.70	EA	3.6	EA	10
3″ × 0.375 WT CS BW	1.50	EA	7.2	EA	11
Handle 3″ 300# control valves—boiler	2.70	EA	2.4	EA	6
3″ × 0.375 WT CS BW	1.50	EA	4.8	EA	7
Handle 2″ 300# control valves—boiler	1.00	EA	2.4	EA	2
2″ × 0.375 WT CS BW	1.10	EA	4.8	EA	5
Handle 4″ 300# flanged flow elements	3.60	EA	7.2	EA	26
4″ 300# Bolt up	1.60	EA	14.4	EA	23
Handle 10″ × 0.562 pipe	1.10	LF	192.0	LF	211
Handle 8″ × 0.562 pipe	0.88	LF	192.0	LF	169
Handle 4″ × 0.375 pipe	0.28	LF	270.0	LF	76
Handle 3″ × 0.375 pipe	0.21	LF	192.0	LF	40
Handle 2″ × 0.375 pipe	0.18	LF	192.0	LF	35

11.8   Waste heat boiler-pressure parts and piping installation man hours

	Actual BM	Estimated BM
Install steam and mud drums	1245	1245
Generating bank tubes	4354	4354
Ground assembly—headers, furnace, and water wall panels	3230	3230
Erect headers and wall panels, fit and weld water wall panels	5291	5291
Fit and weld water wall tubes; 2-31/32″ OD × 0.203″ WT, BW (TIG)	8295	8295
Erect, fit, and weld primary/secondary headers and superheater elements	1817	1817
Burner system	492	492
Sootblowers	1280	1280
Erect and install downcomers and code piping	1811	1811
Waste heat boiler-pressure parts and piping installation man-hours	**27,815**	**27,815**

11.9   Comparison method—Air cooled condenser upper structure a-frame structural

Estimator is using comparison to estimate the erection of A-frame structure in a solar power plant.

11.9.1   Air cooled condenser installation field erection estimate

Description	MH	Qty	Unit	Qty	Unit	IW
Upper structure A-frame steelwork						**294.7**
Erect and temporarily bolt structural steel	2.50	8.52	Ton	8.52	Ton	21.3
Seal plate	2.00	20.00	EA	20.00	EA	40.0
Access door frame	10.00	1.00	EA	1.00	EA	10.0
Cut sheeting to size	0.002	626.00	SF	626.00	SF	1.3
Place sheeting	0.23	420.00	SF	420.00	SF	94.7
Fix sheeting with flashing and caulk	0.05	626.00	SF	626.00	SF	31.3
Tack or screw sheeting	1.00	25.00	EA	25.00	EA	25.0
Install access door	10.00	1.00	EA	1.00	EA	10.0
Pipe hanger with turnbuckle and clevis	10.00	2.00	EA	2.00	EA	20.0
Install steam saddles	10.00	3.00	EA	3.00	EA	30.0
Bolt connection	0.15	74.00	EA	74.00	EA	11.1

11.9.2   Air cooled condenser installation increased quantities estimate

Description	MH	Qty	Unit	Qty	Unit	IW
Upper structure A-frame steelwork						**325.4**
Erect and temporarily bolt structural steel	2.50	9.80	Ton	9.80	Ton	24.5
Seal plate	2.00	20.00	EA	20.00	EA	40.0
Access door frame	10.00	1.00	EA	1.00	EA	10.0
Cut sheeting to size	0.002	719.90	SF	719.90	SF	1.4
Place sheeting	0.23	483.00	SF	483.00	SF	108.9
Fix sheeting with flashing and caulk	0.05	719.90	SF	719.90	SF	36.0
Tack or screw sheeting	1.00	28.75	EA	28.75	EA	28.8
Install access door	10.00	1.00	EA	1.00	EA	10.0
Pipe hanger with turnbuckle and clevis	10.00	2.30	EA	2.30	EA	23.0
Install steam saddles	10.00	3.00	EA	3.00	EA	30.0
Bolt connection	0.15	85.10	EA	85.10	EA	12.8

11.9.3 Air cooled condenser installation decreased quantities estimate

Description	MH	Qty	Unit	Qty	Unit	IW
Upper structure A-frame steelwork						**267.0**
Erect and temporarily bolt structural steel	2.50	7.25	Ton	7.25	Ton	18.1
Seal plate	2.00	20.00	EA	20.00	EA	40.0
Access door frame	10.00	1.00	EA	1.00	EA	10.0
Cut sheeting to size	0.002	532.10	SF	532.10	SF	1.1
Place sheeting	0.23	357.00	SF	357.00	SF	80.5
Fix sheeting with flashing and caulk	0.05	532.10	SF	532.10	SF	26.6
Tack or screw sheeting	1.00	21.25	EA	21.25	EA	21.3
Install access door	10.00	1.00	EA	1.00	EA	10.0
Pipe hanger with turnbuckle and clevis	10.00	2.00	EA	2.00	EA	20.0
Install steam saddles	10.00	3.00	EA	3.00	EA	30.0
Bolt connection	0.15	62.90	EA	62.90	EA	9.4

11.10 Comparison and graphic analysis of data

Comparison of estimated direct craft man-hours		Man-hours
Equipment quantities are decreased; estimated man-hours are decreased	1	267
MHc = field erection man-hours based on historical data	2	295
Equipment quantity is increased; estimated man-hours are increased	3	325

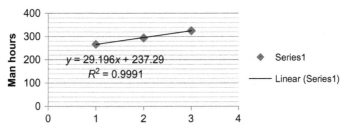

FIG. 11.10.1 Comparison of estimate man-hours.

Estimate is based on historical data from field erection of A-frame structure in a solar power plant (Fig. 11.10.1), and the scope and quantity differences can be identified and the impacts estimated; therefore,

$$MH_\beta \leq MH_c \leq MH_a$$

The proposed unit is based on the estimator's quantity takeoff, and erection quantities are either (+ or −) 15% direct proportion (straight-line graph); comparison method quantities are directly proportional to estimate MHs. Whenever one variable increases or decreases, the other increases or decreases and vice versa.

Chapter 12

Statistical applications to construction

12.1 Section introduction

Statistics is a body of methods enabling us to draw reasonable conclusions from data. Statistics is divided into two general types, descriptive statistics and statistical inference. With descriptive statistics, we summarize data and make calculations and tables or graphs that can be comprehended easily. Statistical inference involves drawing conclusions from the data. In this section, we consider practical ways to enhance man-hour analysis by graphic and analytic techniques. Statistical methods and indexing are considered with the intention to point out the connection to construction. This section provides the reader the knowledge to use construction statistics to collect, analyze, forecast, and use learning curves and time series to validate data and prepare detailed accurate estimates and bid proposals. The section includes practical examples of statistical applications and methods to help the reader understand the importance of man-hour analysis and estimating. To get the most benefits from the statistical applications, the reader should understand exponents, logarithms, and simple algebraic manipulations. It also helps, but is not required, to understand regression analysis. Readers can enter data into the Excel spreadsheets to get the results they need.

Foreman's report and man-hour analysis

Consider an example for the field installation of an air-cooled condenser. Data are collected in the field for the condensate headers. The foreman reports the descriptions of the task, elapsed time, and man-hours completed. The data are examined for consistency, completeness, and accuracy. The report is then compiled for analysis to verify the historical data.

12.2 Analysis of Foreman's report

See Tables 12.2.1 and 12.2.2.

Industrial Process Plant Construction Estimating and Man-Hour Analysis.
https://doi.org/10.1016/B978-0-12-818648-0.00012-0

TABLE 12.2.1 Spreadsheet example for the analysis for welding the condensate headers

Project: Air-cooled condenser—condensate headers			
Foreman: John Smith:		Date: 0000 June 00	
	Craft: Boilermaker		
Cost code	Phase code description	MH	Qty
Condensate headers			
000000	16″ Schedule 10 BW CS	12.00	2.00
000000	18″ Schedule 10 BW CS	13.50	4.00

TABLE 12.2.2 Condensate Headers-air cooled condenser

Phase code description	MH	Qty	Bins
Condensate headers			
16″ Schedule 10 BW CS	12.00	2.00	4.00
18″ Schedule 10 BW CS	13.50	4.00	2.00

Graphic analysis of data

Use Excel's chart capabilities to plot the graph for the Condensate Headers.

To use the Excel chart capabilities, highlight the range F35-G37, and select **Insert, Scatter Chart, Chart Elements, Axis, Axis Titles, Chart Title, Gridlines, Legend, Trendline, and More Options: Display equation on chart and Display R-squared value on chart** (Fig. 12.2.1).

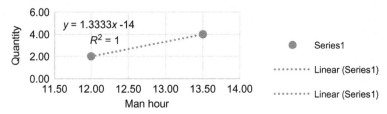

FIG. 12.2.1 Weld condensate headers.

The coefficient of determination, R^2, is exactly $+1$ and indicates a positive fit.

All data points lie exactly on the straight line. The relationship between X and Y variables is such that as X increases, Y also increases.

Histogram

Use Excel's chart capabilities to plot the graph for the histogram.

To use the Excel chart capabilities, highlight range D77-E79, and select **Insert, Charts, Insert Scatter Chart, Insert Combo Chart, Create Custom Combo Chart, Cluster Column, and Chart Elements: Axes, Axes Titles, Chart Title, Gridlines, and Data Table**

Bin	Frequency
2.00	1
4.00	2
More	1

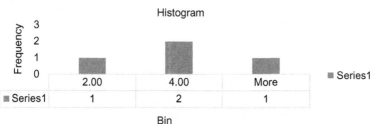

Graphical analysis of process piping and equipment…Validate direct craft man-hour tables

The practical examples that follow for process piping and equipment use graphic analysis to plot the data. Plotting the data is useful to see and understand the relationship between two-variable data and allows results to be displayed in pictorial form. This provides insight into the data set to help with testing assumptions, model selection, regression model validation, relationship identification, factor effect determination, and outlier detection. Plotting of data allows the estimator to check assumptions in statistical models and communicate the results of an analysis.

12.3 Graphical analysis of process piping

Graphs

A graph is a pictorial illustration of the relationship between variables.

Assume for historical man-hour data that x is the independent variable and y is the dependent variable:

Straight-line graph

$$y = a + bx; \quad Y = a + (y - y1)/(x - x1)(x)$$

where y is the dependent variable;

a is the intercept value along the y axis at $x = 0$;
b is the slope, or the length of the rise divided by the length of the run, $b = (y - y1)/(x - x1)$;
and x is the independent or control variable.

Example 12.3.1

Plot a graph for field bolt up of flanged joints by weight class (Table 12.3.1)
  Work scope: Field bolt up flanged joint
  Facility—solar power plant
  Data for input:
  (x): 3, 4, 6, 8, 10, 12, 14, 16, 18, 20, and 24
  (y): 1.20, 1.60, 2.40, 3.20, 4.00, 4.80, 5.60, 6.40, 7.20, 8.00, and 9.60

TABLE 12.3.1 Field bolt up of flanged joints

x	y
Pipe	Man-hour per joint
Size	150#/300#
3	1.20
4	1.60
6	2.40
8	3.20
10	4.00
12	4.80
14	5.60
16	6.40
18	7.20
20	8.00
24	9.60

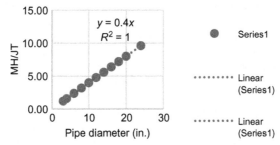

FIG. 12.3.1 Field bolt up flanged joint.

The coefficient of determination, R^2, is exactly $+1$ and indicates a positive fit (Fig. 12.3.1).

All data points lie exactly on the straight line. The relationship between X and Y variables is such that as X increases, Y also increases.

Example 12.3.2

Plot a graph for hydrotest (Table 12.3.2)

    Work scope: Hydrostatic testing

    Facility—Industrial plant

    Data for input:

    X: 2, 2.5, 3, 4, 6, 8, 10, 12, 14, 16, 18, 20, and 24

    y: 0.022, 0.021, 0.025, 0.034, 0.050, 0.067, 0.084, 0.101, 0.118, 0.134, 0.151, 0.168, and 0.202

TABLE 12.3.2 Hydrotest pipe

x	y
Pipe	Man-hours per lineal foot
Size	0.375″ or less
2	0.022
2.5	0.021
3	0.025
4	0.034
6	0.050
8	0.067
10	0.084
12	0.101
14	0.118
16	0.134
18	0.151
20	0.168
24	0.202

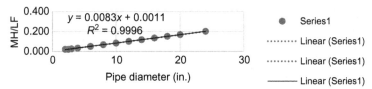

FIG. 12.3.2 Hydrotest pipe.

The coefficient of determination, R^2, is exactly $+1$ and indicates a positive fit (Fig. 12.3.2).

All data points lie exactly on the straight line. The relationship between X and Y variables is such that as X increases, Y also increases.

Example 12.3.3

Plot a graph for handle and install pipe, cast iron soil-no hub (Table 12.3.3)

　　Work scope: Handle and install CI soil pipe, no hub

　　Facility—Industrial plant (underground drainage piping)

　　Data for input:

　　x: 4, 6, 8, 10, 12, 14, 16, 18, 20, and 24

　　y: 0.10, 0.15, 0.20, 0.25, 0.30, 0.35, 0.40, 0.45, 0.50, and 0.60

TABLE 12.3.3 Install underground drainage pipe

x	y
Pipe size	Pipe set and align
4	0.10
6	0.15
8	0.20
10	0.25
12	0.30
14	0.35
16	0.40
18	0.45
20	0.50
24	0.60

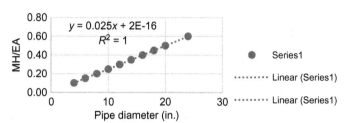

FIG. 12.3.3 Install underground drainage pipe.

The coefficient of determination, R^2, is exactly +1 and indicates a positive fit (Fig. 12.3.3).

All data points lie exactly on the straight line. The relationship between X and Y variables is such that as X increases, Y also increases.

Example 12.3.4

Plot a graph for handle valves by weight class (Table 12.3.4)

Work scope: Handle valves by weight class

Facility—diesel power plant

Data for input:

x: 2, 2.5, 3, 4, 6, 8, 10, 12, 14, 16, 18, 20, and 24

y: 1.80, 1.25, 2.70, 3.60, 5.40, 7.20, 9.00, 10.80, 12.60, 14.40, 16.20, 18.00, and 21.60

TABLE 12.3.4 Handle valves by weight class

x	y
Pipe	MH/EA
Size	600#/900#
2	1.80
2.5	2.25
3	2.70
4	3.60
6	5.40
8	7.20
10	9.00
12	10.80
14	12.60
16	14.40
18	16.20
20	18.00
24	21.60

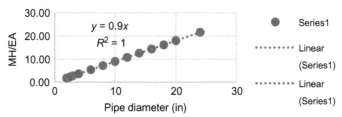

FIG. 12.3.4 Handle valves by weight class.

The coefficient of determination, R^2, is exactly $+1$ and indicates a positive fit (Fig. 12.3.4).

All data points lie exactly on the straight line. The relationship between X and Y variables is such that as X increases, Y also increases.

12.4 Graphical analysis for equipment

Example 12.4.1

Plot a graph for burner system (Table 12.4.1)
  Work scope: Waste heat boiler burner system
  Facility—waste heat boiler
  Data for input:
  x: 280, 90, and 60
  y: 4.7, 1.8, and 1.0

TABLE 12.4.1 Burner system

Burner system	MH	Quantity
	x	x
Burner and wind box	280	4.7
Fan with drive	80	1.8
Silencer	60	1.0

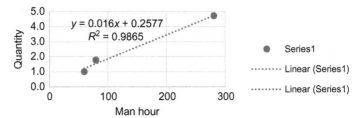

FIG. 12.4.1 Waste heat boiler burner system.

The coefficient of determination is $R^2 = 0.9865$, and the correlation coefficient, $R = 0.9932$, is a strong indication of correlation (Fig. 12.4.1). The relationship between X and Y variables are such that as X increases, Y increases.

Example 12.4.2

Plot a graph for FD fan (Table 12.4.2)
  Work scope: STG—install FD fan
  Facility—coal-fired power plant
  Data for input:
  x: 140, 40, and 60
  y: 2, 0.5, and 1

TABLE 12.4.2 Install FD fan

FD fan	x	y
	MH	Quantity
Set fan and seal weld housing	140	2
Coupling alignment	40	0.5
Install damper	60	1

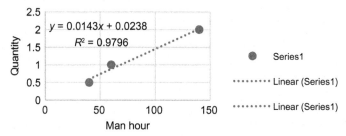

FIG. 12.4.2 Install FD fan.

The coefficient of determination is $R^2 = 0.9796$, and the correlation coefficient, $R = 0.9897$ is a strong indication of correlation (Fig. 12.4.2). The relationship between X and Y variables is such that as X increases, Y increases.

Example 12.4.3

Engines and generators (B2)
  Work Scope: Set engine, couplings, and generator
  Facility—diesel power plant
  Data for input:
  x: 14, 280, and 280
  y: 2792, 924, and 924

Engines, couplings, and generators (B2)	Quantity	MH
Engine generator set 296962#	14	2792
Spring element	280	924
Anchoring plate with AB	280	924

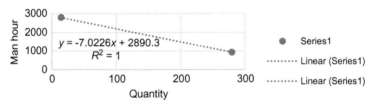

FIG. 12.4.3 Engine, coupling, and generator.

The coefficient of determination, R^2, is exactly +1 and indicates a positive fit (Fig. 12.4.3).
All data points lie exactly on the straight line. The relationship between X and Y variables is such that as X increases, Y also increases.

Example 12.4.4

Install pipe modules
  Work scope: Install pipe modules (B3)
  Facility—diesel power plant
  Data for input:
  x: 14, 112, and 224
  y: 480, 600, and 840

	Quantity	MH
Installation of pipe modules (B3) estimate	x	y
Pipe module	14	480
Bellows	112	600
Bellows (bolt up)	224	840

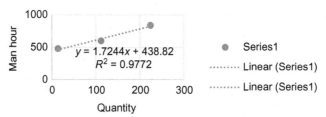

FIG. 12.4.4   Install pipe modules.

The coefficient of determination is $R^2 = 0.9772$, and the correlation coefficient, $R = -0.9885$, is a strong indication of correlation (Fig. 12.4.4). The relationship between X and Y variables is such that as X increases, Y increases.

12.5   Method of least squares for equipment

Least-squares and regression

The practical examples for process piping and equipment that follow use regression models. The least-squares regression model is used to help understand and explain relationships that exist among variables; they are also used to forecast actual outcomes. The reader will learn how least-squares models are derived and use Excel templates to implement them.

Example 12.5.1

Facility—diesel power plant (Table 12.5.1)

Work scope: Set engine, couplings, and generator

Engines, couplings, and generators (B2)
Engine generator set 296962#
Spring element
Anchoring plate with AB

Data for input: Man-hours for field erection of engine, couplings, and generator
Quantity (y): R_1 = 14, 280, and 280
Man-hour (x): R_2 = 2792, 924, and 924

TABLE 12.5.1 Linear regression: fitting a straight line

Description	Quantity	MH
	Y	X
Engines, couplings, and generators (B2)		
Engine generator set 296962#	14	2792
Spring element	280	924
Anchoring plate with AB	280	924

COVAR (R_1, R_2)	-110419.56
VARP (R_2)	15723.56
SLOPE (R_1, R_2)	-7.02
INTERCEPT (R_1, R_2)	2890.32

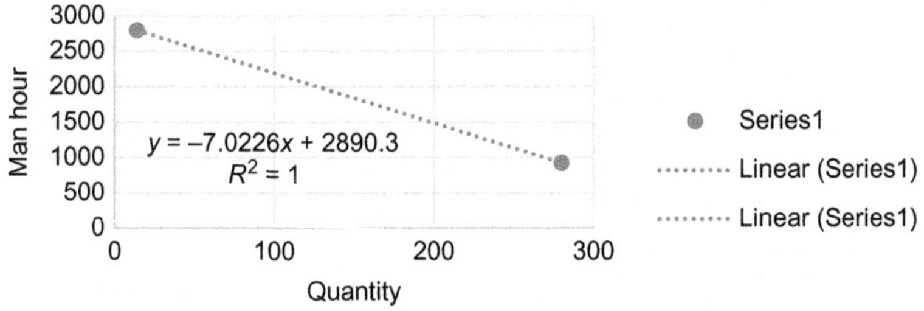

FIG. 12.5.1 Set engine, couplings, and generator.

CORREL (R_1, R_2) = correlation coefficient -1.0000
CORREL $(R_1, R_2)^{\wedge}2$ = coefficient determination 1.0000
The coefficient of determination, R^2, is exactly $+1$ and indicates a positive fit (Fig. 12.5.1).

All data points lie exactly on the straight line. The relationship between X and Y variables is such that as X increases, Y also decreases.

Example 12.5.2
Coal-fired power plant (Table 12.5.2)
　Install FD fan

FD fan
Set fan and seal weld housing
Coupling alignment
Install damper

Scope of field work for field assemble and erect FD fan
Data for input: Man-hours for field assembly and erection of FD fan.
Quantity (y): $R_1 = 2$, 0.5, and 1
Man-hour (x): $R_2 = 140$, 40, and 60

TABLE 12.5.2 Linear regression: fitting a straight line

	Man-hour	Quantity
Description	X	Y
FD fan		
Set fan and seal weld housing	140	2
Coupling alignment	40	0.5
Install damper	60	1

COVAR (R_1, R_2) 26.67
VARP (R_2) 1866.67
SLOPE (R_1, R_2) 0.01
INTERCEPT (R_1, R_2) 0.02

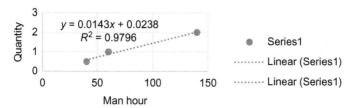

$y = 0.0143x + 0.0238$
$R^2 = 0.9796$

FIG. 12.5.2 Field assemble and erect FD fan.

CORREL (R_1, R_2) = correlation coefficient   1.0000
CORREL $(R_1, R_2)\hat{}2$ = coefficient determination   1.0000
The coefficient of determination is $R^2 = 0.9796$, and the correlation coefficient, $R = 0.9897$, is a strong indication of correlation (Fig. 12.5.2). 98.9% of the total variation on Y can be explained by the linear relationship between X and Y (described by the regression equation, $Y = 0.0143x + 0.0238$). The relationship between X and Y variables is such that as X increases, Y increases.

Example 12.5.3

Coal-fired power plant—spool duct (Table 12.5.3)

Scope of field work for assemble and field erect—spool duct
Spool duct
Casing panel
Casing panel
Roof casing with tube bundle assembly

Data for input: Man-hours for spool duct
Quantity (y): R_1 = 307.4, 175.4, and 175.4
Man-hour (x): R_2 = 107.6, 61.4, and 61.4

TABLE 12.5.3 Linear regression: fitting a straight line

	Man-hour	Quantity
Description	X	Y
Spool duct		
Casing panel	107.6	307.4
Casing panel	61.4	175.4
Roof casing with tube bundle assembly	61.4	175.4

COVAR (R_1, R_2)	1355.20
VARP (R_2)	474.32
SLOPE (R_1, R_2)	2.86

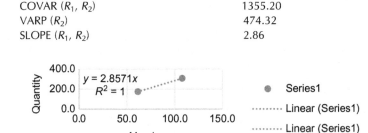

FIG. 12.5.3   Assemble and field erect—spool duct.

INTERCEPT (R_1, R_2)   0.00
CORREL (R_1, R_2) = correlation coefficient   1.0000
CORREL (R_1, R_2)^2 = coefficient determination   1.0000

The coefficient of determination, R^2, is exactly +1 and indicates a positive fit (Fig. 12.5.3).

All data points lie exactly on the straight line. The relationship between X and Y variables is such that as X increases, Y also decreases.

12.6 Method of least squares for process piping

Example 12.6.1

Handle and install PVC pipe (Table 12.6.1)
   Facility—Industrial plant
   Data for input: Man-hours PVC piping
   Man-hour (y): R_1 = 0.12, 0.12, 0.14, 0.18, 0.28, 0.37, 0.46, and 0.55
   Pipe size, inches (x): R_2 = 2, 2.5, 3, 4, 6, 8, 10, and 12
   Man-hour per lineal feet

TABLE 12.6.1 Handle and install PVC piping

X	Y
Pipe	
Size	MH/LF
2	0.12
2.5	0.12
3	0.14
4	0.18
6	0.28
8	0.37
10	0.46
12	0.55

COVAR (R_1, R_2)	0.55
VARP (R_2)	0.02
SLOPE (R_1, R_2)	0.0451
INTERCEPT (R_1, R_2)	0.0084

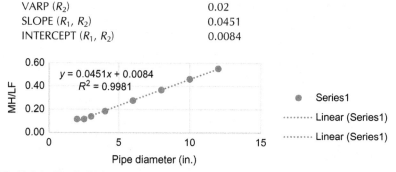

FIG. 12.6.1 Handle PVC piping.

The coefficient of determination, R^2, is exactly +1 and indicates a positive fit (Fig. 12.6.1).

All data points lie exactly on the straight line. The relationship between X and Y variables is such that as X increases, Y also increases.

Example 12.6.2

Diesel power plant—handle and install pipe, carbon steel, and welded joint (Table 12.6.2 and Fig. 12.6.2)

Facility—diesel power plant

Data for input: man-hour field handle S10S and lighter stainless steel pipe

Man-hour (y): R_1 = 0.15, 0.15, 0.18, 0.24, 0.36, 0.48, 0.60, 0.72, 0.84, 0.96, 1.08, 1.20, and 1.44

Pipe size, inches (x): R_2 = 2, 2.5, 3, 4, 6, 8, 10, 12, 14, 16, 18, 20, and 24

Man-hour per LF

TABLE 12.6.2 Handle pipe—0.375″ or less WT

X	Y
Pipe	
size	MH/LF
2	0.15
2.5	0.15
3	0.18
4	0.24
6	0.36
8	0.48
10	0.60
12	0.72
14	0.84
16	0.96
18	1.08
20	1.20
24	1.44

COVAR (R_1, R_2)	2.91
VARP (R_2)	0.17
SLOPE (R_1, R_2)	0.0596
INTERCEPT (R_1, R_2)	0.0067

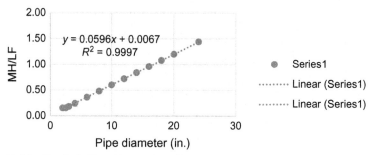

FIG. 12.6.2 Handle S10S and lighter stainless steel pipe.

The coefficient of determination, R^2 is exactly +1 and indicates a positive fit All data points lie exactly on the straight line. The relationship between X and Y variables is such that as X increases, Y also increases.

12.7 Simple moving average forecast

Example 12.7.1

Calculate the forecast values for the time series using a simple moving average with $m = 3$ (Table 12.7.1)

Data for input: Man-hours for the field erection of coal-fired power plant— weld casing seams

Year (t): 2005, 2006, 2007, 2008, 2009, 2010, 2011, and 2012

Average man-hour (y): 1053, 988, 1020, 950, 1061, 1177, 892, and 1089

TABLE 12.7.1 Simple moving average forecast

Year	Average Man-hours			
t	y	*Predict*	(e)	e^2
2005	1053			
2006	988			
2007	1020			
2008	950	1020.2	70.2	4932.7
2009	1061	984.9	75.7	5722.9
2010	1177	1005.3	172.0	29566.8
2011	832	1118.9	287.2	82478.1
2012	1089	1004.4	84.6	7152.1
			MAE	MSE
			137.9	25970.5

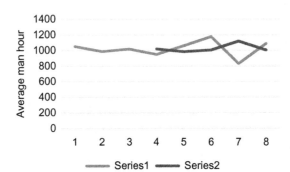

FIG. 12.7.1 Simple moving average coal-fired power plant—weld casing seams.

The plot of forecast values (predict in red) smoothens out the plot of y values (in blue) (Fig. 12.7.1).

The higher the value of *m*, the more smoothing that occurs.

Example 12.7.2

Using Example 1.3 (simple moving average); exponential smoothing with *a* = 0.2 (Table 12.7.2)

Data for input: Man-hours for the field erection of coal-fired power plant—weld casing seams

Year (*t*): 2005, 2006, 2007, 2008, 2009, 2010, 2011, and 2012

Average man-hour (*y*): 1053, 988, 1020, 950, 1061, 1177, 892, and 1089

TABLE 12.7.2 Simple exponential smoothing forecast

Year		Average man-hours		
t	y	*Predict*	e	e^2
2005	1053	1053.0		
2006	988	1053.0	-65.0	4225.0
2007	1020	1040.0	-20.3	412.1
2008	950	1035.9	-85.9	7385.7
2009	1061	1018.8	41.7	1742.9
2010	1177	1027.1	150.1	22529.5
2011	832	1057.1	-225.5	50832.8
2012	1089	1012.0	77.0	5924.5
Predict		8296.9		
	Alpha		MAE	MSE
	0.2		-18.3	13293.2

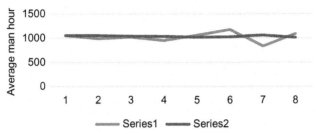

FIG. 12.7.2 Exponential smoothing—coal-fired power plant weld casing seams.

The plot of forecast values (predict in red) smoothens out the plot of y values (in blue) (Fig. 12.7.2).

The lower the value of a, the more smoothing that occurs.

12.8 Application of the learning curve in construction

When a task of work is repeated without interruption, by experienced craft, the repetitive task requires less time and effort. Estimators in construction can apply the theory to productivity and future bidding of similar work. The "learning curve" principle can be applied to industrial construction and is useful for estimating work that is comparable, but work scope quantities may significantly differ and for forecasting output, time, and man-hours. The power equation models the "learning" concept and is used to estimate construction projects. If we use the comparison method for repeat projects, then it takes less time (man-hours) to erect what has been previously erected, which is the learning process. The principle can be applied to industrial construction and modeled by using the "U learning model," power equation, and regression analysis.

12.8.1 U learning model

Learning curves are mathematical models used to estimate efficiencies gained when an activity is repeated. The use of learning curves is to estimate the labor hours in construction when the scope of work is repeated. Learning effects are greatest when the erection process is manual.

Unit (U) model

Used for comparing specific units of production and to fit a U curve to historical data.

The U model is based on the "power law" exponential equation, $y = ax^{\wedge}b$, where a and b are constants. If $b = 1$, the equation is a straight line passing through the origin, with a slope a.

Define the U model: $Hn = H1\ (n^\wedge b)$ where

$Hn =$ hours required for the nth unit of production
$H1 =$ hours required for the 1st unit
$b =$ natural slope of learning curve, illustrating if learning is rapid or slow

If $H1$ and b are known and the unit of production, the estimate of the hours for unit n can be calculated.

Example: Historical data collected from first unit on project 1 is 23,351 h; assume $b = -0.15$, how many hours will the fourth unit require?

$$H4 = H1\ (n^\wedge b) = (23,351) \times (4)^\wedge(-0.15) = 18,967$$

System for measuring slope

The system for measuring slope, that is, rate of learning, measures rate of learning on a scale of 0—100, in percentage.

Slope of $100\% =$ no learning, and $0\% =$ infinity rate of learning.

In practice, effective range of industrial learning is 70%–100%

Natural slope b is defined by the formula $S = 10^\wedge b \log (2) +2$ logarithm to base 10.

Example: If $S = 100\%$ (no learning), $b = \log (100/100)/\log (2) = 0$ $b = 0$ substitute into $Hn = H1\ n^\wedge b$ results in $Hn = H1$ for any value of n (no learning)

Example, if $b = -0.2$, the value of S is?
$S = 10^\wedge(-0.2) \log (2) +2 = 87\%$

Prediction for the total hours for a "block" of production for waste heat boiler erection

Define man-hours for a block of erection as the total man-hours required to erect all units from unit M to another unit $N, N > M$.

TM, N is defined as

$$TM,N = H1\left[M^\wedge b + (M+1)^\wedge b + (M+2)^\wedge b + \cdots + N^\wedge b\right]$$

Approximation formula:

$$TM,N = [H1/(1+b)]\left[(N+0.5)^\wedge(1+b) - (M-0.5)^\wedge(1+b)\right]$$

Example 12.8.1

Historical data for the erection of four waste heat boiler units at project 1

Unit	Man-hour
1	18,967
2	16,813
3	15,513
4	15,357
Total project man-hours	66,650
Y bar =	16,662.5

Analysis of the historical data, for Waste Heat Boiler erection at project 1, unit 1, spent 18,967 man-hours. Estimate the total man-hours to erect the four units, units 1–4, if the learning curve has a slope of 85%.

$$TM, N = [H1/(1+b)] \left[(N+0.5)^{(1+b)} - (M-0.5)^{(1+b)} \right]$$

Value for b: b = log (S/100)/log (2) where S is slope convert to decimal
b = log(85/100)log (2) = log (0.85)log (2) = −0.02125
then, $1 + b = 1 - 0.02125 = 0.9787$
Compute $H1$ using $Hn = Hm (n/m)^{\wedge}b$; $H1$ = (18,967) (1/4) $^{\wedge}$ (−0.02125) = 19,534
Calculate $TM, N = [H1/(1 + b)][(N + 0.5)^{\wedge}(1 + b) - (M - 0.5)^{\wedge}(1 + b)]$, with $N = 4$ and $M = 1$
TM, N = (19,534)[(4.5$^{\wedge}$(0.9787)) − ((0.5$^{\wedge}$0.9787))] = (19,534)(3.8507) = 75,219 man-hours
Actual historical man-hours to install the four units is 66650; results in a 11.39% saving due to learning.

12.8.2   Linear regression—Fitting U model to unit historical data for waste heat boiler erection

Where

$$y = ax^{\wedge}b$$

y = hours required for the nth unit of production
a = hours required for the first unit
b = natural slope

The power function $y = ax^{\wedge}b$ is transformed from a curved line on arithmetic scales to a straight line on log-log scales; let

$$y = \log y$$

$$a = \log a$$

$$x = \log x$$

taking logarithms of both sides, $\log y = \log a + x \log b$, appears like, $y = a + bx$ (Table 12.8.1).

TABLE 12.8.1 Regression analysis of illustrative man-hour data to find intercept a and slope b

Units (x)		Hours (Y)	
1		189.7	
2		168.1	
3		155.1	
4		153.6	
Unit	MH	X = log x	Y = log y
X	Y		
1	189.7	0.0000	2.2780
2	168.1	0.3010	2.2256
3	155.1	0.4771	2.1907
4	153.6	0.6021	2.1863
		1.3802	8.8806

Given: $y = ax^{\wedge}b$ transformed to $\log y = \log a + b \log x$
which is of the form $y = a + bx$.

Let $y = \log y$, $a = \log$ an intercept, $b = b$ slope, and $n =$ sample size.

Linear regression is a method for fitting linear equations of the form $y = a + bx$ to a set of x and y data pairs.

Example 12.8.2

Historical data for project 1; four Waste Heat Boiler Units have been erected with no loss of learning between units. Fit U model to the following historical data: hours = $Y/100$ (Table 12.8.2).

TABLE 12.8.2 Logarithms for x and y

X = log X	Y = log y
0.0000	2.2840
0.3010	2.2324
0.4771	2.1980
0.6021	2.1937

COVAR (R_1, R_2)	-0.0081
VARP (R_2)	0.051123554
SLOPE (R_1, R_2)	-0.1579
INTERCEPT (R_1, R_2)	2.2815

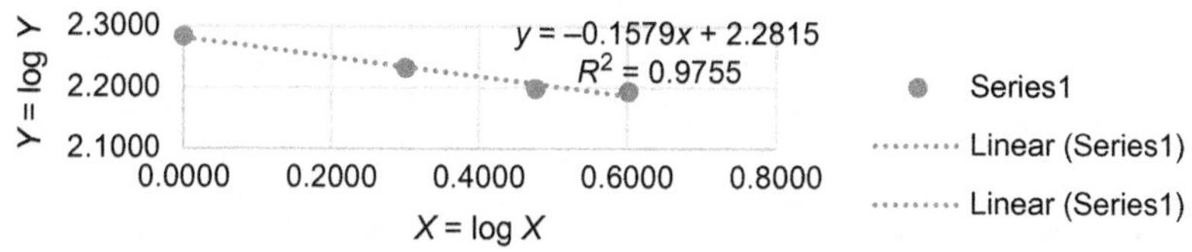

FIG. 12.8.1 Learning U model, $\log y = \log a + b \log x$.

CORREL (R_1, R_2) = correlation coefficient -0.9877
CORREL $(R_1, R_2)^\wedge 2$ = coefficient determination 0.9755
The coefficient of determination is $R^2 = 0.9755$, and the correlation coefficient, $R = -0.9877$, is a strong indicator of correlation (Fig. 12.8.1). The relationship between X and Y variables is such that as X increases, Y increases.

12.9 Risk

Risk is inherent to any construction project. Construction projects can be extremely complex and fraught with uncertainty. Risk analysis is appropriate whenever it is possible to estimate the probability to deal effectively with uncertainty and unexpected events and to achieve project success.

Quantitative risk analysis can be applied to industrial construction and is used to estimate the frequency of risk and the magnitude of their consequences. The application of quantitative risk analysis allows construction project exposure to be modeled and quantifies the probability of occurrence of the identified risk factors and their impact.

12.9.1 Expected-value method

The expected-value method incorporates the effect of risk on outcomes by using a weighted average. Each outcome is multiplied by the probability that the outcome will occur. This sum of products for each outcome is called an expected value. Mathematically, for the discrete case, if X denotes a discrete random variable that can assume the values $X1, X2, \ldots X_i$ with respective probabilities $p1, p2, \ldots p_i$ where $p1 + p2 + \ldots p_i = 1$, the mathematical expectation of X or simply the expectation of X, denoted by $E(X)$, is defined as

$$E(X) = p1X1 + p2X2 + \cdots + p_iX_i = \sum pjXj = \sum pX$$

where $E(X)$ = expected value of the estimate for event i
pj = probability that X takes on value Xj, $0 \le Pj\ (Xj) \le 1$
Xj = event

The pj represents the independent probabilities that their associative Xj will occur with $\Sigma pj = 1$. The expected-value method exposes the degree of risk when reporting information in the estimating process.

Application of the expected value method

Consider the following application of the expected-value method for the construction of an industrial process plant.

The plant will be built in the winter, and the probability is 30% due to excess snow that will delay the mechanical construction for two weeks and cost the project $75,000. Market research indicates there is a 5% probability that the cost of construction material will save the project $55,000. There is a 10% labor productivity factor that will impact the piping installation and cost the project $300,000.

To use the **Math Formulas**, go to quick access toolbar; select **Math & Trig,** and then, select **SUMPRODUCT.**

SUMPRODUCT = Returns the sum of the products of corresponding ranges or arrays.

Arrays 1, 2, and 3 are 2–255 arrays for which you want to multiply and then add components. All arrays must have the same dimensions.

Math & Trig functions are used to calculate values for data.

To use **SUMPRODUCT (array 1, array 2, and array 3),** highlight the array of values for (array 1), highlight the array of values (array R_2), and highlight the array of values (array 3).

Data for input: risk 1, 2, and 3		
Risk	p	X
1	0.30	-$75,000
2	0.05	$55,000
3	0.10	-$300,000

Calculate the expected value for each risk: $E(X) = \sum pX$

Risk			
1	Weather	$E(X)$ = SUMPRODUCT (H71,I71)	−$22,500
2	Material	$E(X)$ = SUMPRODUCT (H72,I72)	$2,750
3	Productivity	$E(X)$ = SUMPRODUCT (H73,I73)	−$30,000
	Project expected value	$E(X)$ = K71 + K72 + K73	−$49,750
	If all risk occur, the project would lose	−$49,750	

12.9.2   Range method

The range method has three estimates, lowest, most likely, and highest for each major cost element. This forms the basis for range estimating.

The mean and variance for each of the three single cost elements are calculated as

$$E(C_i) = (L + 4M + H)/6$$

$$\mathrm{var}(C_i) = ((H - L)/6)^{\wedge}2$$

where $E(C_i)$ = expected cost of distribution i, i = 1, 2, ..., n
L = lowest cost or best-case estimate of cost distribution
M = modal value or most likely estimate of cost distribution
H = highest cost, or worst-case estimate of cost distribution
$\mathrm{var}(C_i)$ = variance of cost distribution i, I = 1, 2, ..., n, dollars^2

The elements are assumed to be independent of each other and are added; then, the new distribution of the total cost is approximately normal. This follows from the central limit theorem. The mean of the sum is the sum of the individual means, and the variance is the sum of the variances:

$$E(Cr) = E(C_1) + E(C_2) + \cdots + E(C_n)$$
$$\mathrm{var}\,(Cr) = \mathrm{var}\,(C_1) + \mathrm{var}\,(C_2) + \cdots + \mathrm{var}\,(C_n)$$

where $E(Cr)$ = expected total cost of independent subdistributions i
$\mathrm{var}\,(Cr)$ = variance of total cost of independent subdistributions i

The probability is calculated using

$$Z = UL - E(Cr)/[\mathrm{var}(Cr)]^1/2$$

where Z = value of the standard normal distribution, Appendix A
UL = upper limit of cost, arbitrarily selected

12.9.3 Application of the range method for equipment

12.9.3.1 Illustration: SCR—Exhaust stack silencer

Data for input (Table 12.9.1):

TABLE 12.9.1 Calculation of expected cost and variance for the installation of exhaust stack silencer probability that the cost will exceed $26,400

L $	M $	H $
4652	4786	4912
4626	4682	4828
4592	4723	5232
6372	5652	5942
6002	6123	6180

Element	L $	M $	H $	E(Cr), $	Var (Cr), $
Support brackets	4652	4786	4912	4784.69	1877.8
Baffles	4626	4682	4828	4696.91	1133.4
Wall bracket	4592	4723	5232	4785.91	11377.8
Baffle guide	6372	5652	5942	5820.24	5136.1
Perforated liner	6002	6123	6180	6112.24	880.1
				$26,200	$20,405

$$PR[\text{mean cost} > E(Cr)] = 0.5 - P(Z)$$

$$Z = UL - E(Cr)/[\text{var}(Cr)]^{1/2}$$

where Z = value of the standard normal distribution, Appendix A
UL = upper limit of cost, arbitrarily selected

Data for input:

UL = $26,400
$E(Cr)$ = $26,200
Var (Cr) = $20,405
Z = 1.40
$P(Z)$; Appendix A; 1.40 = 0.4192
then, $P(\text{Cost} > \$26,400)$ = 8.08%

12.9.4   Contingency

Contingency is a cost element that makes allowance for the unknown risk associated with a project. Contingency cost estimates are assembled when minimum information is provided in the bid documents and where prior experience and data are lacking. Projects requiring excessive research, development, and design are the best candidates for contingency.

Contingency cost estimating is an application of the range method. Three values for each construction task, lowest, L; most likely, M; and the highest, H, are estimated. Values of the upper limit, UL, are estimated and cover the potential range of overrun or underrun of the project. Estimate values depend on the risk involved in construction.

12.9.5   Application of the range method to contingency

Assume the expected cost in Table 12.9.2, $E(Cr) = \$26,200$ and the variance var $(Cr) = \$20,495$, is to be analyzed for contingency.

TABLE 12.9.2 Calculation of expected cost and variance for the installation of exhaust stack silencer

Data for input		
L $	M $	H $
4652	4786	4912
4626	4682	4828
4592	4723	5232
6372	5652	5942
6002	6123	6180

Element	L $	M $	H $	E(Cr), $	Var (Cr), $
Support brackets	4652	4786	4912	4784.69	1877.78
Baffles	4626	4682	4828	4696.91	1133.44
Wall bracket	4592	4723	5232	4785.91	11377.78
Baffle guide	6372	5652	5942	5820.24	5136.11
Perforated liner	6002	6123	6180	6112.24	880.11
				$26,200	$20,405

Probability that the cost will exceed $26,400

$$PR\,[\text{mean cost} > E(Cr)] = 0.5 - P(Z)$$

$$Z = UL - E(Cr)/[\text{var}(Cr)]^1/_2$$

Z = value of the standard normal distribution, Appendix A
UL = upper limit of cost, arbitrarily selected

Data for input (Table 12.9.3):

$UL = \$26,400$
$E(Cr) = \$26,200$
$Var\ (Cr) = \$20,405$
$Z = 1.40$
$P(Z)$; Appendix A; $1.40 = 0.4192$
then, $P(\text{Cost} > \$26,400) = 8.08\%$

TABLE 12.9.3 Calculation of probability given that a cost upper limit exceeds the mean value

UL, $	Z	Appendix	P(cost ≥ UL)	Formula
$25,900.00	(2.10)	0.4821	0.98	0.5 + 0.4772
$26,000.00	(1.40)	0.4192	0.92	0.5 + 0.4162
$26,100.00	(0.70)	0.2580	0.76	0.5 + 0.2580
$26,200.00	0.00	0.0000	0.50	0.5 + 0.0000
$26,250.00	0.35	0.1368	0.36	0.5 − 0.1406
$26,290.00	0.63	0.2357	0.26	0.5 − 0.2422
$26,330.00	0.91	0.3186	0.18	0.5 − 0.3238
$26,390.00	1.33	0.4082	0.09	0.5 − 0.4066
$26,450.00	1.75	0.4599	0.04	0.5 − 0.4641
$26,500.00	2.10	0.4821	0.02	0.5 − 0.4772
$26,550.00	2.45	0.4929	0.01	0.5 − 0.4918

Risk graph of contingency (Fig. 12.9.1)

Probabilities	
Range, $	Probability
$25,900.00	0.98
$26,000.00	0.92
$26,100.00	0.76
$26,200.00	0.50
$26,250.00	0.36
$26,290.00	0.26
$26,330.00	0.18
$26,390.00	0.09
$26,450.00	0.04
$26,500.00	0.02
$26,550.00	0.01

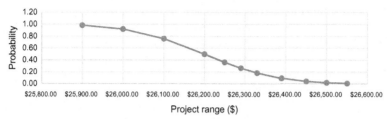

FIG. 12.9.1 Risk graph of contingency for exhaust stack silencer.

Observe Fig. 12.9.1; the probability the expected value exceeds the expected cost is 50%.

Therefore, the values less than $26,200 are greater than 50%.

From the analysis, the estimator can estimate the contingency from the risk graph.

12.10 Bid assurance

The contractor needs to bid high enough to make profit but low enough to get the job.

The "optimum bid" or the "best bid" will result in a successful bid. When the bid leads to winning a job, then there is an opportunity to verify the estimate's accuracy, reliability, and quality. The purpose of this section is to provide methods for optimizing the bid and regulate cost. This defines bid assurance.

12.10.1 Expected profit

It is important the contractor understand the bidding process. For example, the contractor will bid high and win no jobs or bid low and get many jobs with no profit. The higher the bid, the lower the chance of success. The lower bid has better chances of success but more chance of loss. The difference between the bid price and the cost depends on the contractor's need for work, minimum acceptable markup, and the maximum contractor markup. Markup is defined as the difference between the bid price and the estimated cost. Expected profit is defined as

Profit: $P = Bp - Ec$
Expected profit: $Ep = Pr^*(B - Ec)$

where

Bp = bid price
Ec = estimated cost
Pr = probability of event $(Bp - Ec)$, $0 \leq Pr \leq 1$

Example—application of expected profit

Contractor is bidding on a job he estimates will cost $620,000 to complete. The contractor feels he has some knowledge how the competitor is likely to bid. In the following table, the contractor has estimated the subjective probabilities of being the lowest bidder:

Data for input (Table 12.10.1):

TABLE 12.10.1 Expected profit for project with a $620,000 cost estimate

Bp, $	Pr(Bp < next lowest bid)	Ec = estimated cost
630,000	1.00	620,000
620,000	0.95	620,000
630,000	0.85	620,000
645,000	0.45	620,000
650,000	0.15	620,000
660,000	0.00	620,000

Bp, $	Pr(Bp < next lowest bid)	Ec = estimated cost	Expected profit, $Pr(Bp − Ec)
590,000	1.00	620,000	−$30,000
620,000	0.95	620,000	$0
630,000	0.85	620,000	$8,500
645,000	0.45	620,000	$11,250
650,000	0.15	620,000	$4,500
660,000	0.00	620,000	$0

Probabilities for various bids

Bid, B, $	P(B < next lowest bid)
$590,000	1.00
$620,000	0.96
$630,000	0.80
$645,000	0.45
$650,000	0.15
$660,000	0.00

FIG. 12.10.1 Graph of probability of bid less then lowest bid.

The bid includes profit and cost. Contractor long-term profit is a function of successful bids and the profit and the capture rate. Winning and losing bids are a function related to cost estimating (Fig. 12.10.1).

12.10.2 Capture rate

The capture rate is defined as

capture rate $= (Cs)/(C_i) \times 100$, percent; $Cs =$ cost estimates are successful; $C_i =$ total cost estimate.

If the capture rate declines, then profit is reduced and vice versa. Contractor can operate successfully with a 10% capture rate. The capture rate needs to be modified to dollars that win to dollars that were bid. Efficiency is output/input that is less than 1. The output/input factor divides the new estimate to indicate an "adjusted estimate"; efficiency $= \sum a / \sum e$.

12.11 Application of optimal bidding strategy

Contractor is bidding on a project that he expects will cost $400,000 to complete.

Contractor has excess capacity and can take on another project. Several other contractors are bidding the project. Contractor has bid against several of the bidders in the past and feels he has data on how they are likely to bid.

In the following table, contractor has estimated the subjective probabilities of winning for any bid he makes.

Contract award will be to the lowest bidder. The higher the contractor bids, the more his profit but less chance of award. Contractor must find optimal bid for success.

Contractor's profit if he is low bidder and his bid is less the $400,000 cost.

Calculation for expected profit:

If contractor bids $575,000, then the profit is $575,000–$400,000 = $175,000. If the bid is not successful, the profit is zero.

Hence, E(profit) = (given bid profit $\times$ probability of bid success) + (0 $\times$ probability of losing) = profit if bid wins $\times$ probability of winning (Tables 12.11.1 and Fig. 12.11.1).

TABLE 12.11.1 Expected profit

Bid, $	(1) Profit if bid wins (bid−$400,000)	(2) Probability of winning	(3) Expected profit (1) × (2)
350,000	−$50,000	1.00	-$50,000
375,000	−$25,000	0.95	−$23,750
400,000	$0	0.90	$0
425,000	$25,000	0.80	$20,000
450,000	$50,000	0.60	$30,000
475,000	$75,000	0.35	$26,250
500,000	$100,000	0.20	$20,000
525,000	$125,000	0.10	$12,500
550,000	$150,000	0.05	$7,500
575,000	$175,000	0.00	$0

To maximize expected profit, the contractor should bid $450,000.

Bid, $	Probabilities Probability of winning
$350,000	1.00
$375,000	0.95
$400,000	0.90
$425,000	0.80
$450,000	0.60
$475,000	0.35
$500,000	0.20
$525,000	0.10
$550,000	0.05
$575,000	0.00

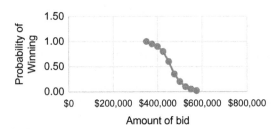

FIG. 12.11.1 Graph of optimal bidding strategy.

Appendix A

Areas of standard normal distribution

Areas under the normal curve

z	0.00	0.01	0.02	0.03	0.04	0.05	0.06	0.07	0.08	0.09
0.0	0.0000	0.0040	0.0080	0.0120	0.0159	0.0199	0.0239	0.0279	0.0319	0.0359
0.3	0.1179	0.1217	0.1255	0.1293	0.1331	**0.1368**	0.1406	0.1443	0.1480	0.1517
0.6	0.2257	0.2291	0.2324	**0.2357**	0.2389	0.2422	0.2454	0.2486	0.2518	0.2549
0.7	**0.2580**	0.2611	0.2642	0.2673	0.2704	0.2734	0.2764	0.2794	0.2823	0.2852
0.9	0.3159	**0.3186**	0.3212	0.3238	0.3264	0.3289	0.3315	0.3340	0.3365	0.3389
1.3	0.4032	0.4049	0.4066	**0.4082**	0.4099	0.4115	0.4131	0.4147	0.4162	0.4177
1.4	**0.4192**	0.4207	0.4222	0.4236	0.4251	0.4265	0.4279	0.4292	0.4306	0.4319
1.7	0.4554	0.4564	0.4573	0.4582	0.4591	**0.4599**	0.4608	0.4616	0.4625	0.4633
2.1	**0.4821**	0.4826	0.4830	0.4834	0.4838	0.4842	0.4846	0.4850	0.4854	0.4857
2.4	0.4918	0.4920	0.4922	0.4925	0.4927	**0.4929**	0.4931	0.4932	0.4934	0.4936

This table gives the probability of a random variable falling in the range $z = 0$ to $z = z$.

Appendix B

Statistical and mathematical formulas

Statistical formulas for the mean, variance, and standard deviation

Mean: $\bar{y} = y1 + y2 + \cdots + yn$; $\Sigma\ y/n$
Variance: $S^2 = (y1 - \overline{Y})^2 + (Y2 + \overline{Y}^2) + \cdots + (Yn - \overline{Y})^2/n - 1$

$$s^2 = \Sigma(yi - \bar{y})^2/(n-1)$$

Standard-deviation: $S = [(y1 - \overline{Y})^2 + Y2 + \overline{Y}^2) + \cdots + (Yn - \overline{Y})^2/n - 1]^{1/2}$

$$s = \left[\Sigma(yi - \bar{y})^2/(n-1)\right]^{1/2}$$

Straight line graph: handle and install large bore standard pipe

$$y = a + bx;\quad Y = a + (y - y1)/(x - x1)(x)$$

where

y = dependent variable
a = intercept value along the y-axis at $x = 0$
b = slope or the length of the rise divided by the length of the run;
$$b = (y - y1)/(x - x1)$$
x = independent or control variable

Mathematical expectation: $E(X) = p1X1 + p2X2 + \cdots + \rho kXk = \sum \rho X$
Normal distribution: $Y = 1/(\sigma(2pi)^{\wedge}1/2)e^{\wedge} - 1/2(X - \mu)^2/\sigma^2$
  where μ = mean, σ = standard deviation, pi = 3.1416..., and e = 2.71828....
Standard form: $Y = 1/(2pi^{\wedge}1/2)e^{\wedge} - 1/2(z^2)$
  z is normally distributed with mean 0 and variance 1
Central limit theorem: $W_i = ((\bar{x})_i - \mu)/(\sigma/k^{\wedge}1/2)$
  N(0,1) in the limit as k approaches infinity

Method of least squares

Least-square line

The least-square line approximating the set of points (x1, y1), (x2, y2) ... (xn, yn) has the equation

$$y = bx + a$$

where

b = the slope of the line
a = y-intercept

The best fit line for the points (x1, y1), (x2, y2) ... (xn, yn) is given by

$$y - \overline{y} = b(x - \overline{x})$$

where the slope is

$$b = \Sigma(xi - \overline{x})(yi - \overline{y})/\Sigma(xi - \overline{x})^2$$

and the y-intercept is

$$a = \overline{y} - b\overline{x}$$

Formula correlation coefficient r:

$$r = n(\Sigma XY) - (\Sigma X)(\Sigma Y)/\left[(n(\Sigma X^2 - (\Sigma X)^2)\right]\left[n\left(\Sigma y^2 - (\Sigma Y)^2\right)\right]^{1}/_{2}$$

Define the U model: Hn = H1 (n$^{\wedge}$b)
where

Hn = hours required for the nth unit of production
H1 = hours required for the first unit

Natural slope b is defined by the formula $S = 10^{\wedge}b\log(2)+2$ logarithm to base 10

Prediction for the total hours for a "block" of production

Define man-hours for a block of erection as the total man-hours required to erect all units from unit M to another unit N, N > M.
  TM, N is defined as

$$TM,N = H1\left[M^{\wedge}b + (M+1)^{\wedge}b + (M+2)^{\wedge}b + \cdots + N^{\wedge}b\right]$$

Approximation formula:

$$TM,N = [H1/(1+b)]\left[(N+0.5)^{\wedge}(1+b) - (M-0.5)^{\wedge}(1+b)\right]$$

Linear regression—Fitting U model to unit historical data

$$y = ax^{\wedge}b$$

where

y = hours required for the nth unit of production
a = hours required for the first unit
b = natural slope

The power function $y = ax^{\wedge}b$ is transformed from a curved line on arithmetic scales to a straight line on log-log scales; let

$$y = \log y$$

$$a = \log a$$

$$x = \log x$$

Taking logarithms of both sides, $\log y = \log a + x \log b$ appears like $y = a + bx$.

Calculating sample size

Sample size: absolute precision

$$n = z^{2*}p^*(1 - p)/e^2$$

where

z = z value (1.96 for 95% confidence level)
p = percentage expressed as a decimal (0.2)
e = acceptable error percentage as a decimal (.05 = ±5%)

Calculating error limits for a sample size

Given sample size is calculated by

$$n = \left(z^{2*}p^*(1 - p)\right)/e^2$$

Determine the limit of error, e

$$e = z^*(p^*(1 - p)/n)^{\wedge}1/2$$

Work sampling method

Model: $Hs = (N_i)\ (Ht)\ (RF)\ (1 + PF\&D)/N$
   where

Hs = standard man-hours per task
N_i = observation of event i
Ht = total man-hours worked during sample study
RF = rating factor
$PF\&D$ = personal, fatigue, and delay allowance
N = number of random observations during sample study

Chapter 12 Analysis of risk probability in construction

Expected value method

If X denotes a discrete random variable that can assume the values X1, X2, ..., X_i with respective probabilities p1, p2, ..., p_i where $p1 + p2 + \cdots p_i = 1$, the mathematical expectation of X or simply the expectation of X denoted by E(X) is defined as

$$E(X) = p1\,X1 + p2\,X2 + \cdots + p_iX_i = \sum pj\,Xj = \sum pX$$

where

$E(X)$=expected value of the estimate for event i
pj=probability that X takes on value Xj, $0 <= Pj\,(Xj) <= 1$
Xj=event

Range method

The mean and variance for each of the three single cost elements are calculated as

$$E(C_i) = (L + 4M + H)/6$$

$$var(C_i) = ((H - L)/6)^{\wedge}2$$

where

$E(C_i)$=expected cost of distribution i, $i = 1, 2, ..., n$
L=lowest cost or best-case estimate of cost distribution
M=modal value or most likely estimate of cost distribution
H=highest cost or worst-case estimate of cost distribution
$var(C_i)$=variance of cost distribution i, $I = 1, 2, ..., n$, dollars^2

The mean of the sum is the sum of the individual means, and the variance is the sum of the variances:

$$E(Cr) = E(C1) + E(C2) + \cdots + E(Cn)$$

$$var\,(Cr) = var\,(C1) + var\,(C2) + \cdots + var\,(Cn)$$

where

$E(Cr)$=expected total cost of independent subdistributions i
var (Cr)=variance of total cost of independent subdistributions i

The probability is calculated using

$$Z = UL - E(Cr)/[var(Cr)]^{\frac{1}{2}}$$

where

Z=value of the standard normal distribution, Appendix A
UL=upper limit of cost, arbitrarily selected

Expected profit is defined as

$$\text{profit} = \text{Bp} - \text{Ce}$$

$$\text{expected profit} = P^* \left(\text{Bp} - \text{Ce}\right)$$

where

Bp = bid price
Ce = estimated cost
P = probability of event (Bp − Ce), $0 <= \text{Prob} <= 1$

Capture rate

The capture rate is defined as

$$\text{capture rate} = \left(\text{Cs}\right) / \left(C_i\right)^* 100, \text{percent}$$

where Cs = cost estimates that are successful

Moving averages. Smoothing of time series

Given a set of numbers

$$Y1, Y2, \ldots$$

define a moving average of order N to be given by the sequence of arithmetic means:

$$Y1 + Y2 + \cdots + Yn/n, Y2 + Y3 + \cdots + yn/n, Y3 + Y4 + \cdots + yn/n, \ldots$$

The sums in the numerators are moving totals of order n

Estimation of moving averages

The average of n most recent observations, computed at time t, is given by

$$\text{Ma} = \text{yt} + \text{yt} - 1 + \cdots + \text{yt} - n + 1/n$$

where

Ma = moving average of response variable
y = data, labor, cost, price, etc.
t = unit of time, years, months, etc.
n = denominator of group of time units

Exponential smoothing

Formula:

$$\left(D^*S\right) + \left(F^* \left(1 - s\right)\right)$$

where

D = most recent period's demand
S = smoothing factor in decimal form
F = most recent period's forecast

Cost index

The current cost is found by using the formula

$$C = H\,(Ic/Ih)$$

where

C = current or future cost
H = historical or past cost
Ic = index corresponds to current or future time period
Ih = index corresponds to historical or past time period

Appendix C

Excel functions and mathematical functions

Excel functions

Graphic analysis of data

Use Excel's chart capabilities to plot the graphic straight line given by the equation $y = a + bx$

To use the Excel chart capabilities, highlight the range x:y, and select **insert,** and select from **Charts, Scatter;** go to quick access bar, and select from **Chart Tools, design,** and from **Chart Layouts,** select **Layout 9.**

Excel functions

Excel statistical functions for forecasting the value of y for any x. Thus, a and b can be calculated in Excel. Where R1 = the array of y values and R2 = the array of x values

b = SLOPE (R1, R2)=COVAR (R1, R2)/VARP (R2)

a = INTERCEPT (R1, R2)=AVERAGE (R1) – b * Average (R2)

SLOPE (R1, R2) = slope of regression line

INTERCEPT (R1, R2) = y-intercept of the regression line

FORECAST (x, R1, R2) calculates the predicted value of y for given value of x. Thus,

FORECAST (x, R1, R2) = a+b * x where a = INTERCEPT (R1, R2) and b = SLOPE (R1, R2)

TREND (R1, R2) = array function that produces an array of predicted y values corresponding to x values stored in array R2, based on the regression line calculated from x values stored in array R2 and y values stored in array R1.

COVAR (R1, R2) = returns covariance, the average of the products of deviations for each data point pair in two data cells

VARP (R2) = calculates variance based on the entire population (ignores logical values and text in the population)

Correlation

CORREL (R1, R2)=correlation coefficient of data in arrays R1 and R2

CORREL (R1, R2)$^\wedge 2$=coefficient determination

The Mean, Variance and Standard Deviation Measures of Central Tendency AVERAGE (number 1, number 2): Returns the average (arithmetic mean) of its arguments, which can be numbers or names, arrays, or references that contain numbers.

VAR (number 1, number 2): estimates variance based on a sample (ignores logical values and text in the sample **Math & Trig function Math Formulas**).

STDEV (number 1, number 2): estimates standard deviation based on a sample (ignores logical values and text in the sample **Math & Trig function Math Formulas**). To use the **Math Formulas**, go to quick access toolbar; select **Math & Trig,** and then, select **SUMPRODUCT SUMPRODUCT**=returns the sum of the products of corresponding ranges or arrays; arrays 1, 2, and 3 are 2–255 arrays for which you want to multiply and then add components. All arrays must have the same dimensions.

LOG=returns the logarithm of a number to the base you want the logarithm.

Number is the positive real number for which you want the logarithm.

SQRT=returns the square root of a number. Number is the number for which you want the square root.

Appendix D

Area and volume formulas

Formulas—areas and volumes

Square: area $=(\text{edge})^2$; $A = a^2$

Rectangle: base $\times$ altitude; $A = ba$

Right triangle: area $= 1/2$ base x altitude; $A = 1/2\ ba$

Pythagorean theorem:

$(\text{Hypotenuse})^2 +$ sum of squares of two legs of right triangle

$$c^2 = a^2 + b^2; \quad a = \left[c^2 - b^2\right]^{\wedge}1/2$$

Oblique triangle: area $= 1/2$ base $\times$ altitude; $A = 1/2\ bh$

$A = [s(s-a)(s-b)(s-c)]^{\wedge}1/2$, where $s = (a+b+c)/2$

Parallelogram: opposite sides are parallel; $A = bh$

Trapezoid: one pair of opposite sides parallel

$A = 1/2$ sum of bases x altitude; $A = 1/2\ (a+b)\ h$

Circle: circumference $= 2(\text{pi})\ (\text{radius}) = (\text{pi})\ (\text{diameter}$: $C = 2(\text{pi})R = (\text{pi})D$

Area $= (\text{pi})\ (\text{radius})^2 = (\text{pi})/4)\ (\text{diameter})^2$: $A = (\text{pi})R^2 = (\text{pi}/4)D^2$

Sector of circle: area $= 1/2$ radius $\times$ arc; $A = 1/2\ Rc = 1/2\ R^2$ angle

Segment of circle: area (segment) $=$ area (sector) $-$ area (triangle)

$A = Rc - 1/2\ ba$

Ellipse: area $= (\text{pi})ab$

Parabolic segment: area $= 2/3\ ld$

Right circular cone: $V = (\text{pi})r^2h$; $A =$ side area $+$ base area

$$A = (\text{pi})r\left[r + \left(r^2 + h^2\right)^{\wedge}1/2\right]$$

Right circular cylinder: $V = (\text{pi})r^2h = (\text{pi})d^2h/4$

$A =$ side area $+$ end areas $= 2\ (\text{pi})r\ (h+r)$

Appendix E

Standard to metric

Lengths

Metric conversion

1 centimeter = 10 millimeters; 1 cm = 10 mm
1 meter = 100 centimeters; 1 m = 100 cm
Standard conversions
1 foot = 12 inches; 1 ft = 12 in
1 yard = 3 feet; 1 yd = 3 ft
1 yard = 36 inches; 1 yd = 36 in

Metric-standard conversions

1 millimeter = 0.03937 inches; 1 mm = 0.03937 in
1 centimeter = 0.39370 inches; 1 cm = 0.39370 in
1 meter = 39.39008 inches; 1 m = 39.37008 in
1 meter = 3.28084 feet; 1 m = 3.28084 ft
1 meter = 1.093.6133 yards; 1 m = 1.0993.6133 yd

Standard-metric conversions

1 inch = 2.54 centimeters; 1 in = 2.54 cm
1 foot = 30.48 centimeters; 1 ft = 30.48 cm
1 yard = 91.44 centimeters; 1 yd = 91.44 cm
1 yard = 0.9144 meters; 1 yd = 0.3144 m

Volumes

Metric conversion

1 cubic centimeter = 1000 cubic millimeters; 1 cu cm = 1000 cu mm
1 cubic meter = 1 million cubic centimeters; 1 cu m = 1,000,000 cu cm

Standard conversions

1 cubic foot = 1728 cubic inches; 1 cu ft = 1728 cu in
1 cubic yard = 46,656 cubic inches; 1 cu yd = 46,656 cu in
1 cubic yard = 27 cubic feet; 1 cu yd = 27 cu ft

Metric-standard conversions

1 cubic centimeter = 0.06102 cubic inches; 1 cu cm = 0.06102 cu in
1 cubic meter = 35.31467 cubic feet; 1 cu m = 35.31467 cu ft
1 cubic meter = 1.30795 cubic yards; 1 cu m = 1.30795 cu yd

Standard-metric conversions

1 cubic inch = 16.38706 cubic centimeters; 1 cu in = 16.38706 cu cm
1 cubic foot = 0.02832 cubic meters; 1 cu ft + 0.02832 cu m
1 cubic yard = 0.76455 cubic meters; 1 cu yd = 0.76455 cu m

Areas

Metric conversion

1 sq centimeter = 100 sq millimeters; 1 sq cm = 100 sq mm
1 sq meter = 10,000 sq centimeters; 1 sq m = 10,000 sq cm

Standard conversions

1 sq foot = 144 sq inches; 1 sq ft = 144 sq in
1 sq yard = 9 sq feet; 1 sq yd = 9 sq ft

Metric-standard conversions

1 sq centimeter = 0.15500 sq inches; 1 sq cm = 0.15500 sq in
1 sq meter = 10.76391 sq feet; 1 sq m = 10.76391 sq ft
1 sq meter = 1.19599 sq yards; 1 sq m = 1.9599 sq yd

Standard-metric conversions

1 sq inch = 6.4516 sq centimeters; 1 sq in = 6.4516 sq cm
1 sq foot = 929.0304 sq centimeters; 1 sq ft = 929.0304 sq cm
1 sq foot = 0.09290 sq meters; 1 sq ft = 0.09290 sq m

Appendix F

Boiler man-hour tables

TABLE F.1 Cutting and milling rates for 0–5000 psi

Size (OD)	Torch cut	Saw or grind	Mill with power tool
0″ < Diameter <= 3″	0.15–0.20	0.35–0.50	0.35–0.50
3″ < Diameter <= 4-1/2″	0.18–0.30	0.375–1.00	0.375–1.00
4-1/2″ < Diameter <= 6-1/2″	0.25–0.60	0.50–3.00	0.50–3.00
6-1/2″ < Diameter <= 35″ with WT from 1/2″ to 5-1/2″	Torch cut		

Wachs Trav-L-Cutter—use
0.033–0.050 MH/inch of circumference for every 1/2″ of wall thickness to be removed
Mill with power tool—use
0.05–0.10 MH/inch of circumference for every 1/2″ of wall thickness to be removed
Saw or grind—use mill with power tool rate
Formula: circumference in inches × rate/inch × no. of passes (1/2″ cut per pass)
Example: 22″ diameter × 2″ wall
[(22″ diameter × 3.1416) × 0.033/inch] (2″ wall 1/2)=9.1 hrs

Circumference	Rate	No. of passes

Rates for carbon steel only, double for all other material
Excludes overhead labor, setup, maintenance, and removal of equipment

TABLE F.2 Expanding rates, MH per tube end

PSI	2″	2-1/2″	3″	4″
160	0.19	0.22	0.44	0.94
200	0.20	0.24	0.45	0.95
300	0.22	0.27	0.47	0.96
400	0.25	0.31	0.49	0.97
500	0.28	0.36	0.50	0.99
600	0.30	0.40	0.53	1.02
700	0.36	0.43	0.59	1.07
800	0.36	0.43	0.66	1.33
900	0.36	0.43	0.72	1.60
1000	0.36	0.43		1.87
1100	0.50			
1200	0.50			
1300	0.50			
1400	0.50			
1500	0.50			
1600	0.50			
1700	0.50			
1800	1.20			
1900	1.20			
2000	1.20			
2100	1.20			

TABLE F.3 Socket and seal welding, MH per weld

| MH per weld size | Seal welding | | Socket welding | |
	Outside header	Inside header	Without stress relieving	With stress relieving
1″ OD tube	0.7	0.9	1.1	1.4
Over 1″ to 1-1/2″ OD tube	0.9	1.1	1.2	1.5
Over 1-1/2″ to 2″ OD tube	1.2	1.3	1.4	1.7

TABLE F.3 Socket and seal welding, MH per weld—cont'd

MH per weld size	Seal welding		Socket welding	
	Outside header	*Inside header*	*Without stress relieving*	*With stress relieving*
Over 2″ to 2-1/2″ OD tube	1.4	1.5	1.7	2.2
Over 2-1/2″ to 3-1/4″ O.D. tube	1.7	1.8	2.1	2.6
Over 3-1/4″ to 4″ OD tube	1.9	2.1		
Over 4″ to 4-1/2″ OD tube	2.1	2.4		
Over 4-1/2″ to 5-1/2″ OD tube	2.3	2.7		
Place and weld				
2-1/2″ HH fitting	2.2			
3-1/4″ HH fitting	2			
4″ HH fitting or blind nipple	2.5			
4-1/2″ HH fitting or blind nipple	2.7			
Radiograph and header end plugs	2.8			

For seal welding 2″ generating tubes inside drums, reexpanding and NDE, use 1.2 MH/JT.

TABLE F.4 Field tube welding, MH/weld

	Tube welding—MH/weld					
	Design pressure (PSI)					
	Up to 500	*501–1000*	*1001–1500*	*1501–2000*	*2001–2500*	*2501–3000*
Tube size (OD)						
1″ < BW <=1-1/2″ TIG	2.4	2.6	2.8	3.0	3.2	3.4
1-1/2″ <BW <=2″ TIG	2.7	2.9	3.4	3.7	4.0	4.3
2″ < BW <=2-1/2″ TIG	3.2	3.3	3.8	4.2	4.2	4.4

Continued

TABLE F.4 Field tube welding, MH/weld—cont'd

| | Tube welding—MH/weld | | | | | |
| | Design pressure (PSI) | | | | | |
	Up to 500	501–1000	1001–1500	1501–2000	2001–2500	2501–3000
2-1/2″ < BW <=3″ TIG	3.7	4.0	4.1	4.8	4.6	4.8
3″ < BW <=3-1/2″ TIG	4.3	4.6	4.8	5.3	5.3	5.6
3-1/2″ < BW <=4″ TIG	4.9	5.2	5.4	5.9	5.9	6.2
4″ < BW <=4-1/2″ TIG	5.5	5.8	6.1	6.5	6.5	6.7
4-1/2″ < BW <=5-1/2″ TIG	6.8	7.1	7.4	7.7	8.0	8.3
4-1/2″ < Ring weld <= 5-1/2″ SMAW PWHT	8.4	9.4	10.3	11.5	12.4	13.6
5-1/2″ < Ring weld <= 6-1/2″ SMAW	5.0	5.8	6.4	6.8	7.0	7.2
5-1/2″ < BW <=6-1/2″ TIG	7.7	8.0	8.5	9.0	9.5	10.0
5-1/2″ < Ring weld <= 6-1/2″ SMAW PWHT	10.2	11.0	12.0	13.8	14.2	15.0

Field welding of tubes in the heat input zones of all boilers of 2000 psi design and over is by the TIG process.
Field welding of tubes by the SMAW process; root pass by the TIG process.

PWHT:

A. carbon steel >3/4″ thick
B. chrome molly steel with carbon content >0.25% and wall thickness > 1/2″
C. croloy materials with more than 3% chromium or diameter >4″ and WT >1/2″

TABLE F.5 Field pipe welding, MH/weld

| | Wall thickness in inches | | | | | | | | | | | |
Diameter (inches)	0.250	0.500	0.750	1.000	1.250	1.500	1.750	2.000	2.250	2.500	2.750	3.000
6	3.0	3.3	7.2	8.7	13.2	15.0	16.2	17.4				
8	4.0	4.4	9.6	11.6	17.6	20.0	21.6	23.2	24.8	28.0		
10	5.0	5.5	12.0	14.5	22.0	25.0	27.0	29.0	31.0	35.0	38.0	43.0
12	6.0	6.6	14.4	17.4	26.4	30.0	32.4	34.8	37.2	42.0	45.6	51.6
14	7.0	7.7	16.8	20.3	30.8	35.0	37.8	40.6	43.4	49.0	53.2	60.2
16	8.0	8.8	19.2	23.2	35.2	40.0	43.2	46.4	49.6	56.0	60.8	68.8
18	9.0	9.9	21.6	26.1	39.6	45.0	48.6	52.2	55.8	63.0	68.4	77.4
20	10.0	11.0	24.0	29.0	44.0	50.0	54.0	58.0	62.0	70.0	76.0	86.0
22	11.0	12.1	26.4	31.9	48.4	55.0	59.4	63.8	68.2	77.0	83.6	94.6
24	12.0	13.2	28.8	34.8	52.8	60.0	64.8	69.6	74.4	84.0	91.2	103.2
26	13.0	14.3	31.2	37.7	57.2	65.0	70.2	75.4	80.6	91.0	98.8	111.8
28	14.0	15.4	33.6	40.6	61.6	70.0	75.6	81.2	86.8	98.0	106.4	120.4
30	15.0	16.5	36.0	43.5	66.0	75.0	81.0	87.0	93.0	105.0	114.0	129.0
32	16.0	17.6	38.4	46.4	70.4	80.0	86.4	92.8	99.2	112.0	121.6	137.6
34	17.0	18.7	40.8	49.3	74.8	85.0	91.8	98.6	105.4	119.0	129.2	146.2
35	17.5	19.3	42.0	50.8	77.0	87.5	94.5	101.5	108.5	122.5	133.0	150.5

PWHT:

A. carbon steel >3/4″ thick
B. chrome moly steel with carbon content >0.25% and wall thickness >1/2″
C. croloy materials with more than 3% chromium or diameter >4″ and WT >1/2″

TABLE F.6 Diamond sootblowers, MH per SB W/PVF

Type of unit	Manual operation	MH	Unit
G-2, G-21, G-9B, A2E	SB	22	EA
	SB W/PVF	44	EA
	Total	66	EA
IR and 1S	SB	20	EA
	SB W/PVF	44	EA
	Total	64	EA
1K, T9, and T11	SB	35	EA
	SB W/PVF	51	EA
	Total	86	EA
G9B, A2E (swinging arm)	SB	18	EA
	SB W/PVF	51	EA
	Total	69	EA
1K, DE2, SE2 (strait line)	SB	100	EA
	SB W/PVF	51	EA
	Total	151	EA
Pressure reducing station		70	EA
Trays and channels for supporting tubing		18	EA
Tubing for automatic sequential air operation		40	EA
Air compressor and/or receiver		18	Ton
Auto sequential panel or air master controller		86	Panel
IK structural supports		6	Blower

For combustion steam and air blowing sootblowers, double PVF man-hours.
For Vulcan sootblowers, add 24 MH per sootblower for scavenger drain PVF.

For removal of sootblower, use 50% of table man-hours.

TABLE F.7 Structural steel

Structural steel and miscellaneous iron erect structural steel		
Main steel	*MH*	*Unit*
Erect structural steel; <=20 tons		
Light—0–19 lb/ft	28.0	Ton
Medium—20–39 lb/ft	24.0	Ton
Heavy—40–79 lb/ft	20.0	Ton
X heavy—80–120 lb/ft	16.0	Ton
Erect structural steel; 20 tons > tons <= 100 tons		
Light—0–19 lb/ft	21.0	Ton
Medium—20–39 lb/ft	18.0	Ton
Heavy—40–79 lb/ft	15.0	Ton
X heavy—80–120 lb/ft	12.0	Ton
Erect structural steel; >100 tons		
Light—0–19 lb/ft	16.8	Ton
Medium—20–39 lb/ft	14.4	Ton
Heavy—40–79 lb/ft	13.2	Ton
X heavy—80–120 lb/ft	11.8	Ton
Platform framing	0.15	SF
Handrail and toeplate	0.25	LF
Floor grating	0.20	SF
Stair treads	0.85	LF
Straight ladder	0.30	LF
Caged ladders	0.35	LF
Ladders and safety cage—KD	1.00	LF
Girths (sidewall support steel for metal siding)	28.00	Ton
Purlins (roof support steel for metal siding)	28.00	Ton
Elevator steel	38.00	Ton

TABLE F.8 Burners

Description	MH	Unit
Circular burners	32.00	Ton
Air jet low No$_\chi$ coal burner	180.00	EA
DRB-4Z low No$_\chi$ coal burner	240.00	EA
XLC low No$_\chi$ oil and gas burner	240.00	EA
Dual zone No$_\chi$ port (over fire air system)	200.00	EA
Intertube burners include blocks, tips, and riffle casings	18.00	Ton
Shop-assembled burners in wind box include welding	24.00	Ton
Automatic lighters include welding	6.00	EA
Oil atomizers	4.00	EA

Glossary

Definitions for productivity analysis

Actual time The time reported for work, which includes delays, idle time, and inefficiency, as well as efficient effort.

Allowance An adjustment for work, which includes delays, idle time, and inefficiency, as well as efficient effort.

Continuous timing A method of time study where the total elapsed time from the start is recorded at the end of each element.

Cycle The total time of elements from start to finish.

Delay allowance One part of the allowance included in the standard time for interruptions or delays beyond the worker's ability to prevent.

Element A subpart of an operation or task separated for timing and analysis; beginning and ending points are described, and the element is the smallest part of an operation observed by time study. The length of the element can vary from minutes, to hours, to days, etc.

Fatigue allowance An allowance based on physiological reduction in ability to do work, sometimes included in the standard time.

Idle time An interval in which the worker, equipment, or both are not performing useful work.

Normal time An element or operation time found by multiplying the average time observed for one or multiple cycles by a rating factor.

Observed time The time observed on the stopwatch/electronic clock or other media and recorded on the time study sheet or media tape during the measurement process.

Person/man-hour A unit of measure representing one person working for 1 hour.

Productivity rate A unit rate of production; the total amount produced in a given period divided by the number of hours or months or years.

Productivity The amount of work performed in a given period. Usually measured in units of work per man-hour or man-month or man-year

Rating factor A means of comparing the performance of the worker under observation by using experience or other benchmarks; additionally, a numerical factor is noted for the elements or cycle; 100% is normal, and rating factors less than or greater than normal indicate slower or faster performance.

Standard time Sums of rated elements that have been increased for allowances.

Task/operation Designated and described work sublet to work measurement, estimating, and reporting.

Piping abbreviations

BOT bottom
BW butt weld
C to C center to center
COL column
CONN connection
CS carbon steel
CV control valve
DIA diameter
EL elevation
ELO elbolet
FAB fabrication
FF flat face
FLG flange
FT feet
FW field weld
ID inside diameter
IN inch
ISO isometric
LB large bore pipe
LG long
LOL latrolet
MATL material
No number
NOM nominal
OD outside diameter
Olet branch connection

PL plate
PS pipe support
PWHT post weld heat treatment
RF raised face
SB small bore pipe
SCD screwed
SCH schedule
SMLS seamless
SOL slip on
SOL sockolet
SS stainless steel
STD standard
STM steam
STR straight
SUPPT support
SW socket weld
TOL threadolet
TOS top of steel or support
TYP typical
VA valve
VERT vertical
WN weld neck
WOL weldolet
WT wall thickness
XS extra strong
XXs double extra strong

Labor factors

Dilution of supervision This occurs when supervision is diverted from productive, planned, and schedules work to analyze and plan contract changes, expedite delayed material, manage added crews, or other changes not in the original work scope and schedule. Dilution is also caused by an increase in manpower, work area, or project size without an increase in supervision.

Errors and omissions Increases in errors and omissions impact on labor productivity because changes are then usually performed on a crash basis, out of sequence, cause dilution of supervision, or any other negative impacts.

Fatigue Fatigue can be caused by prolonged or unusual exertion.

Joint occupancy This occurs when work is scheduled utilizing the same facility or work area that must be shared or occupied by more than one craft and not anticipated in the original bid or plan.

Learning curve When crew turnover causes new workers to be added to a crew or additional manpower is needed within a crew, a period of orientation occurs in order to become familiar with changed conditions. They must then learn work scope, tool locations, work procedures, and so on.

Morale and attitude Sprit of workers based on willingness, confidence, discipline, and cheerfulness to perform work or task can be lowered due to a variety of issues, including increased conflicts, disputes, excessive hazards, overtime, overinspection, multiple contract changes, disruption of work, rhythm, poor site conditions, absenteeism, and unkempt work space.

Overtime Scheduling of extended work days or weeks exceeding a standard 8 hour work day or 40 hour. Work week lowers work output and efficiency, physical fatigue, and poor mental attitude.

Reassignment of manpower When workers are reassigned, they experience unexpected or excessive changes, losses caused by move-on or move-off, reorientation, and other issues that result in a loss of productivity.

Ripple This is caused when changes in other trades' work then affect other works, such as the alteration of schedule.

Stacking of trades This occurs when operations take place within physically limited space with other contractors resulting in congestion of personnel, inability to use or locate tools conveniently, increased loss of tools, additional safety hazards, increased visitors, and prevention of crew size optimum.

Weather/season changes Performing work in a change of season. Temperature zone or climate change resulting in work performed in either very hot or very cold weather, rain or snow, or there are changes in temperature or climate that can impact workers beyond normal conditions.

Index

Note: Page numbers followed by *f* indicate figures, *t* indicate tables, and *b* indicate boxes.

W

Waste heat boiler
  boiler casing, 69
    equipment installation man hours, 78
    estimate data, 77
    field work required, 76
    installation estimate, 77
  equipment installation man-hours, 75
  estimate data
    down comers and code piping sheet 3, 72
    drums, generating tubes, headers, side,
      front and rear wall panels sheet 1, 71
    erect headers and panels, weld water wall
      tubes, burner, soot blower, superheater
      sheet 2, 71
  500,000 lb/h installation estimate
    down comers and code piping sheet 3, 232
    drums, generating tubes, headers, side,
      front and rear panels sheet 1, 230

    erect headers and panels, weld water wall
      tubes, burner, soot blower, superheater
      sheet 2, 231
  installation estimate
    down comers and code piping sheet 3,
      75
    drums, generating tubes, headers, side,
      front and rear panels sheet 1, 73
    erect headers and panels, weld water wall
      tubes, burner, soot blower, superheater
      sheet 2, 74
  pressure parts, 67–68
  structural steel
    equipment installation man hours, 78
    estimate data, 76
    field work required, 76
    installation estimate, 77
  work-field erection, 70
Work sampling method, 273

Printed in the United States
By Bookmasters